AMELIA COUNTY [VIRGINIA] ROAD ORDERS

1735-1753

Virginia Genealogical Society
Richmond, Virginia

Published With Permission from the

Virginia Transportation Research Council
(A Cooperative Organization Sponsored Jointly by the Virginia
Department of Transportation and
the University of Virginia)

HERITAGE BOOKS
2008

HERITAGE BOOKS
AN IMPRINT OF HERITAGE BOOKS, INC.

Books, CDs, and more—Worldwide

For our listing of thousands of titles see our website
at
www.HeritageBooks.com

Published 2008 by
HERITAGE BOOKS, INC.
Publishing Division
100 Railroad Avenue #104
Westminster, Maryland 21157

Copyright © 2002 Virginia Genealogical Society

All rights reserved. No part of this book may be reproduced or transmitted in any form or by any means, electronic or mechanical, including photocopying, recording or by any information storage and retrieval system without written permission from the author, except for the inclusion of brief quotations in a review.

International Standard Book Number: 978-0-7884-3657-4

AMELIA COUNTY ROAD ORDERS 1735-1753

by

Nathaniel Mason Pawlett
Faculty Research Historian

Ann Brush Miller
Senior Research Scientist

and

Kenneth Madison Clark
Research Associate

Virginia Transportation Research Council
(A Cooperative Organization Sponsored Jointly by the Virginia
Department of Transportation and
the University of Virginia)

Charlottesville, Virginia

April 2002
VTRC 02-R14

DISCLAIMER

The contents of this report reflect the views of the authors, who are responsible for the facts and the accuracy of the data presented herein. The contents do not necessarily reflect the official views or policies of the Virginia Department of Transportation, the Commonwealth Transportation Board, or the Federal Highway Administration. This report does not constitute a standard, specification, or regulation.

Copyright 2002 by the Commonwealth of Virginia

HISTORIC ROADS OF VIRGINIA

Louisa County Road Orders, 1742-1748, by Nathaniel Mason Pawlett. 57 pages, indexed, map.

Goochland County Road Orders, 1728-1744, by Nathaniel Mason Pawlett. 120 pages, indexed, map.

Albemarle County Road Orders, 1744-1748, by Nathaniel Mason Pawlett. 57 pages, indexed, map.

The Route of the Three Notch'd Road, by Nathaniel Mason Pawlett and Howard Newlon. 26 pages, illustrated, 2 maps.

An Index to Roads in the Albemarle County Surveyor's Books, 1744-1853, by Nathaniel Mason Pawlett. 10 pages, map.

A Brief History of the Staunton and James River Turnpike, by Douglas Young. 22 pages, illustrated, map.

Albemarle County Road Orders, 1783-1816, by Nathaniel Mason Pawlett. 421 pages, indexed.

A Brief History of Roads in Virginia, 1607-1840, by Nathaniel Mason Pawlett. 41 pages.

A Guide to the Preparation of County Road Histories, by Nathaniel Mason Pawlett. 26 pages, 2 maps.

Early Road Location: Key to Discovering Historic Resources? by Nathaniel Mason Pawlett and K. Edward Lay. 47 pages, illustrated, 3 maps.

Albemarle County Roads, 1725-1816, by Nathaniel Mason Pawlett. 98 pages, illustrated, 8 maps.

"Backsights," A Bibliography, by Nathaniel Mason Pawlett. 29 pages, revised edition.

Orange County Road Orders, 1734-1749, by Ann Brush Miller. 323 pages, indexed, map.

Spotsylvania County Road Orders, 1722-1734, by Nathaniel Mason Pawlett. 159 pages, indexed.

Brunswick County Road Orders, 1732-1749, by Nathaniel Mason Pawlett. 81 pages, indexed.

Orange County Road Orders, 1750-1800, by Ann Brush Miller. 394 pages, indexed, map.

Lunenburg County Road Orders, 1746-1764, by Nathaniel Mason Pawlett and Tyler Jefferson Boyd. 394 pages, indexed.

Culpeper County Road Orders, 1763-1764, by Ann Brush Miller. 22 pages, indexed, map.

Augusta County Road Orders 1745-1769, by Nathaniel Mason Pawlett, Ann Brush Miller, Kenneth Madison Clark and Thomas Llewellyn Samuel, Jr. 270 pages, indexed, map.

Requests for information as to availability and
a current price list should be directed to:

Historic Roads of Virginia
Virginia Transportation Research Council
530 Edgemont Road
Charlottesville, VA 22903

www.virginiadot.org/vtrc/history/roadordr.html

DEDICATION

This volume is dedicated to the memory of

Nathaniel Mason Pawlett (1935-1995)

Faculty Research Historian, Virginia Transportation Research Council

1973-1995

"gentleman and scholar"

FOREWORD

by

Ann Brush Miller

Amelia County Road Orders 1735-1753 is one of several compilations of early Virginia road orders left unfinished by Nathaniel Mason Pawlett, longtime Faculty Research Historian of the Virginia Transportation Research Council, at his death in 1995. Mr. Pawlett began work on the first volumes of published Virginia road orders in the early 1970s, and during the next twenty years produced more than a dozen volumes of road order transcriptions and histories of Virginia roads.

The production of this volume involved the efforts of a number of individuals. Tyler Jefferson Boyd assisted Mr. Pawlett on the initial transcription of the Amelia County road orders. Subsequent completion of the volume was undertaken by Ann B. Miller and Kenneth M. Clark, assisted by Valerie B. Huber.

This volume marks the twentieth entry in the *Historic Roads of Virginia* series, first initiated by the Virginia Transportation Research Council (then the Virginia Highway & Transportation Research Council) in 1973. *Amelia County Road Orders 1735-1753* expands the coverage of the early Southside Virginia transportation records begun in the previously published *Brunswick County Road Orders 1732-1749* and *Lunenburg County Road Orders 1746-1764*.

A NOTE ON THE METHODS, EDITING, AND DATING SYSTEM

by

Nathaniel Mason Pawlett

The road and bridge orders contained in the order books of an early Virginia county are the primary source of information for the study of its roads. When extracted, indexed, and published by the Virginia Transportation Research Council, they greatly facilitate this. All of the early county court order books are in manuscripts, sometimes so damaged and faded as to be almost indecipherable. Usually rendered in the rather ornate script of the time, the phonetic spellings of this period often serve to complicate matters further for the researcher and recorder.

With these road orders available in an indexed and cross-indexed published form, it will be possible to produce chronological chains of road orders illustrating the development of many of the early roads of a vast area from the threshold of settlement through much of the eighteenth century. Immediate corroboration for these chains of road orders will usually be provided by other evidence such as deeds, plats, and the Confederate Engineers maps. Often, in fact, the principal roads will be found to survive in place under their early names.

With regard to the general editorial principles of the project, it has been our perception over the years as the road orders of Louisa, Hanover, Goochland, Albemarle, and other counties have been examined and recorded that road orders themselves are really a variety of "notes," often cryptic, incomplete, or based on assumptions concerning the level of knowledge of the reader. As such, any further abstracting or compression of them would tend to produce "notes" taken from "notes," making them even less comprehensible. The tendency, therefore, has been in the direction of restraint in editing, leaving any conclusions with regard to meaning up to the reader or researcher using these publications. In pursuing this course, we have attempted to present the reader with a typescript text that is as near a type facsimile of the manuscript itself as we can come.

Our objective is to produce a text that conveys as near the precise form of the original as we can, reproducing all the peculiarities of the eighteenth-century orthography. Although some compromises have had to be made because of the modern keyboard, this was really not that difficult a task. Most of their symbols can be accommodated by modern typography, and most abbreviations are fairly clear as to meaning.

Punctuations may appear misleading at times, with unnecessary commas or commas placed where periods should be located; appropriate terminal punctuation is often missing or else takes the form of a symbol such as a long dash, etc. The original capitalization has been retained insofar as it was possible to determine from the original manuscript whether capitals were intended. No capitals have been inserted in place of those originally omitted. The original spelling and syntax have been retained throughout, even including the obvious errors in various places, such as repetitions of words and simple clerical errors. Ampersands have been retained throughout to include such forms as "&c" for "etc." Superscript letters have also been retained where used in y^e, y^t, s^d. The thorn symbol (y), pronounced as "th," has been retained in the aforesaid "y^e," pronounced "the," and "y^{t}" (that). The tailed "p" (resembling a capital "p" with the tail extended into a loop) has also been retained. This symbol has no counterpart in modern typography; given the limitations of the modern keyboard, we have rendered it as a capital "p" (P). This should be taken to mean either "per" (by), "pre," or "pro" (sometimes "par" as in "Pish" for parish) as the context of the order may demand. For damaged and missing portions of the manuscripts, we have used square brackets to denote the [missing], [torn], or [illegible] portions. Because of the large number of ancient forms of spelling, grammar, and syntax, it was deemed impracticable to insert the form *[sic]* after each one to indicate a literal rendering. Therefore, the reader must assume that apparent errors are merely the result of our literal transcription of the road orders, barring the introduction of typographical errors, of course. If, in any case, this appears to present insuperable problems, resort should be made to the original records.

As to dating, most historians and genealogists who have worked with early Virginian records will be aware of the English dating system in use down to 1752. Although there was an eleven-day difference from our calendar in the day of the month, the principal difference lay in the fact that the beginning of the year was dated from March 25 rather than January 1, as was the case from 1752 onward to the present. Thus January, February, and March (to the 25th) were the last three months in a given year, and the new year came in only on March 25.

Early Virginian records usually follow this practice, though in some cases, dates during these three months will be shown in the form 1732/3, showing both the English date and that in use on the Continent, where the year began January 1. For researchers using material with dates in the English style, it is important to remember that under this system (for instance) a man might die in January 1734 yet convey property or serve in public office in June 1734 since, under this system, June came *before* January in a given year.

INTRODUCTION

by

Nathaniel Mason Pawlett

and

Ann Brush Miller

> The roads are under the government of the county courts, subject to be controuled by the general court. They order whenever they think them necessary. The inhabitants of the county are by them laid off into precincts, to each of which they allot a convenient portion of the public roads to be kept in repair. Such bridges as may be built without the assistance of artificers, they are to be built. If the stream be such as to require a bridge of regular workmanship, the court employs workmen to build it, at the expense of the whole county. If it be too great for the county, application is made to the general assembly, who authorize individuals to build it, and to take a fixed toll from all passengers, or give sanction to such other proposition as to them appears reasonable.
>
> —Thomas Jefferson, *Notes on the State of Virginia*, 1781.

The establishment and maintenance of public roads were among the most important functions of the County Court during the colonial period in Virginia. Each road was opened and maintained by an Overseer of the Highways appointed by the Gentlemen Justices yearly. He was usually assigned all the "Labouring Male Titheables" living on or near the road for this purpose. These individuals then furnished all their own tools, wagons, and teams and were required to labor for six days each year on the roads.

Major projects, such as bridges over rivers, demanding considerable expenditure were executed by Commissioners appointed by the Court to select the site and to contract with workmen for the construction. Where bridges connected two counties, a commission was appointed by each and they cooperated in executing the work.

At its creation from portions of Prince George County and Brunswick County in 1735, Amelia County occupied that portion of Southside Virginia between the giant parent county of Brunswick to the south and the Appomattox River on the north and extended westward to the head of that river near what would become Appomattox Court House in 1845. North of the Appomattox River lay another large parent county, Goochland, which, like Brunswick, extended to the Blue Ridge.

In modern terms, at its creation, Amelia County extended through a sizable portion of central Southside Virginia. Stretching from its eastern border with what is now Dinwiddie

County westward to the vicinity of Lynchburg, and bounded on the north by the Appomattox River and on the south by the dividing ridge between the Appomattox and Roanoke river systems, Amelia at its greatest extent included the modern counties of Amelia, Nottoway, and Prince Edward, along with a portion of Appomattox County.

From 1735 to 1754, Amelia was a moderately large parent county, not so large as Brunswick, but occupying a position key to any understanding of the road development of Southside Virginia. The first division of Amelia's territory came in 1754, when Prince Edward County (then containing a portion of Appomattox) was formed from the western half of Amelia. The second and final division of the county came in 1789 with the creation of Nottoway County, by which Amelia County reached its present configuration.

The road orders contained in this volume cover the period from the creation of Amelia County in 1735 through its first division (the creation of Prince Edward County) in 1754. As such, these records are the principal extant evidence concerning the early development of that area of central Southside Virginia through which ran the main roads connecting the frontier areas to the west with the capital at Williamsburg and the budding commercial centers of Richmond and Petersburg.

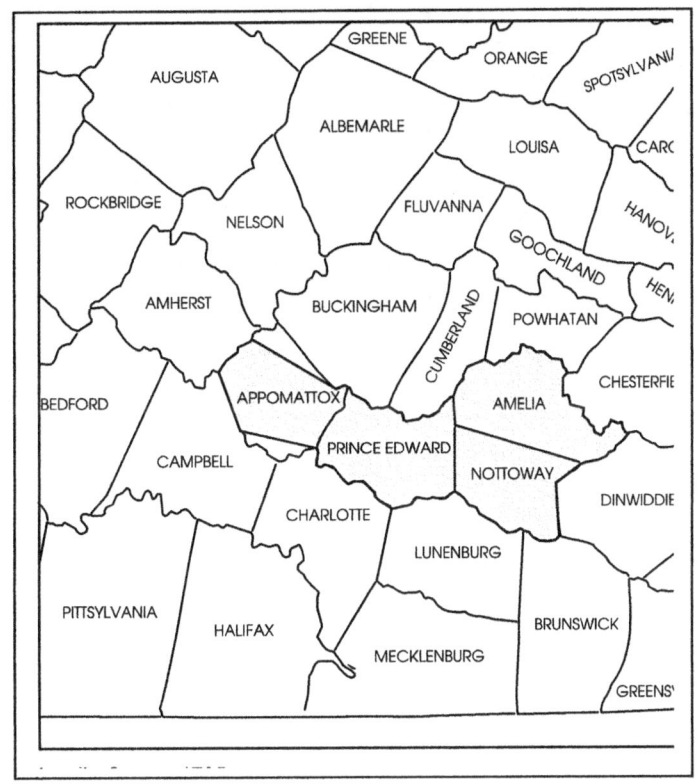

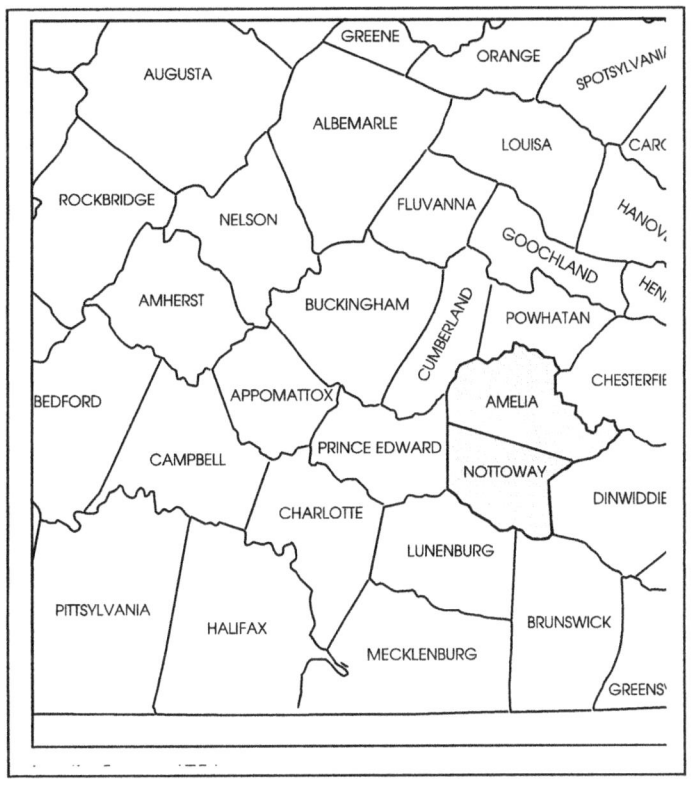

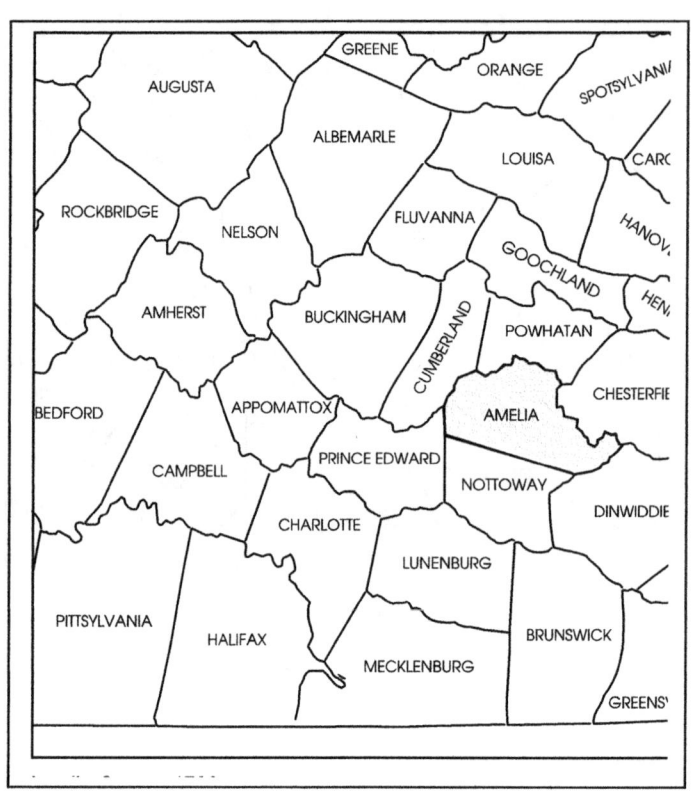

Amelia County, 1735-1753

Amelia County Order Book I

9 May 1735 Old Style, Page 1
Abraham Burton Appointed Surveyor of the Road from Flatt Creek to Sapp[page torn - Sappony(?)] ford and Deep Creek lower bridg

9 May 1735 Old Style, Page 1
William Green Surveyor from y^e. afore Said Bridg to Namoz^n. Ro[page torn - Road(?)]

9 May 1735 Old Style, Page 1
John Forguson Surveyor of the Road from Flatt Creek & down Andersons Road

9 May 1735 Old Style, Page 1
Lewis Tanner Ordered to Clear a Road from the bridg below Mrs Ander[page torn - Anderson?] Quarter to Buckskin

8 August 1735 Old Style, Page 2
Robert Tucker Ju^r Appointed Surveyor of that Road from Namoz^n. bridg to the Swett house Creek

8 August 1735 Old Style, Page 3
John Nance appointed Surveyor from Butterwood to the coming in of the race paths near M^r. Irbys

8 August 1735 Old Style, Page 3
Christopher Robertson Surveyor from Bagleys order to Capt Starks quarter

8 August 1735 Old Style, Page 3
Francis Alan Surveyor of the Road a Cross Beaver pond branch into Andersons road

14 November 1735 Old Style, Page 4
Christopher Hinton Appointed Surveyor of the Road over the Swethouse Creek below Abraham Jones's quarter into the Main road

14 November 1735 Old Style, Page 4
David Lyles Surveyor of a bridle way from the Rattlesnake ford to the Church on Flatt Creek

14 November 1735 Old Style, Page 4
William Clarke Appointed Surveyor of the Road from West Creek to the fork of the road near Capt Peter Jones quarter

14 November 1735 O. S., Page 4
William Crawley Appointed Surveyor of the Road from Cap Peter Jones fork where Clarke leaves off to Wintercomake

14 November 1735 O. S., Page 4
Isaac Hudson appointed Surveyor of the road from Smax Creek to the river and Deep Creek bridg

14 November 1735 O. S., Page 4
Several of the Inhabitants of this County complaining of the hardships they labour under occasioned by the riseing of the water in Appomatox river so as to prevent their crossing the same praying that a boat or some other convenience may be provided for that purpos. The Court conceiveing the same to resonable. do appoint Edward Booker and Samuel Cobbs Gent. to consult with the Gent. of Henrico Court what measures they think proper to be taken

13 February 1735 O. S. Page 6
Bryan Fenell Appointed Surveyor of a Road to be clear'd a Little below John Jacksons to the fork of Little Nottoway

13 February 1735 O. S. Page 7
Robert Taylor Appointed Surveyor from M.r Andersons Bridge to Deep Creek Bridge

13 February 1735 O. S. Page 7
John Benson Appointed to clear a bridle [page torn - way(?)] from West Creek to the Court house

13 February 1735 O. S. Page 7
John Winingham Appointed to Clear a bridle way from West Creek to the Chapell on Nottoway

12 March 1735 O. S. Page 7
Cap.t Richard Jones & M.r Abraham Green are desired to Consult with the Court of Prince George about Repairing Namozain bridge and to agree with Some person to do the same

12 March 1735 O. S., Page 7
Mich.l McDearman Row app.d Surveyor from Dabneys to the Cutt bancks

9 April 1736 O. S., Page 8
Edward Booker Gent desireing the Road may be continued as it now is from the fork of Bookers road to the Church which is accordingly granted

9 April 1736 O. S., Page 8
David Lyles had leive granted him to keep a ferry from his house over Appm.x river. the price for a man & horse Six pence for one person three pence

9 April 1736 O. S., Page 8
Henry Anderson has leive to turn the Road near his house making another as good as this now is

14 May 1736 O. S., Page 9
Edwin Franklin Appointed Surveyor of the Road from the bent Creek into Bookers road

14 May 1736 O. S., Page 10
James Anderson Appointed Surveyor of the high ways where Christ.º Robertson was Surveyor. order'd that the Said Anderson cause the said ways to be cleard according to law

14 May 1736 O. S., Page 10
George Bagley appointed Surveyor of the high ways from the Celler to Dandys Race paths ordered the ways be cleared According to law

13 August 1736 O. S., Page 13
Samuel Pincham is Appointed Surveyor of the Road from Hall Creek to Smax Creek in Room of Abr: Burton Dec[d]

12 November 1736 O. S., Page 15
Thomas Covington Appointed Surveyor in room of Lewis Tanner

10 December 1736 O. S., Page 17
John Forguson Appointed Surveyor of the Road from Flatt Creek to the Court house

10 December 1736 O. S., Page 18
County Levy
For Repairing Namozain bridge … 134

14 January 1736 O. S., Page 19
Richard Booker Gent is desired to Agree with a Workman to build a boat which is to be kept by Thomas Bevill who is to sett over all Persons Inhabiting in this County and all others haveing tithables therein those that have not to pay four pence for Man & Horse. for a Cart Eight pence for a Chair or two wheel Chaise four pence

11 March 1736 O. S., Page 21
Charles Burkes appointed Surveyor of a Road be cleared from the Courthouse to Andersons Road near the race paths Ordered he cause the same to be don and Bridges where they are wanting

11 March 1736 O. S., Page 21
Robert Vaughan Appointed Surveyor of a Road to be clear'd from Tho[s] Bottoms on West Creek to the Old Ponds of Flatt Creek along or Near the old Ridge Path Ordered he cause the same to be don and Bridges made where they are wanting

11 March 1736 O. S., Page 21
William Echols Appointed Surveyor of a Road to be Cleared from the White Oak on Flatt Creek to John Hurts near the Fork of Stocks Creek Ordered he Cause the same to be don and Bridges made where they are wanting

11 March 1736 O. S., Page 22
John Dawson gent is Apointed Surveyor of a Road to be cleared from fflatt Creek to or near the ffork of Saylors Creek and it is Ordered that those persons that shall be thought most convenient by the said Dawson (if they are not employed on other roads) assist in clearing the same.

13 May 1737 O. S., Page 25
Henry Tally is apointed to clear a Road from Deep Creek to this [page torn] House and it is Ordered that the Hands convenient assist in clearing the Same

13 May 1737 O. S., Page 26
William Shannon is apointed Surveyor of the Road in the room of Isaac Hudson.

13 May 1737 O. S., Page 27
Grand Jury Presentments
... the Road from fflat Creek to Mrs Andersons Quarter is out of repair, ...

13 May 1737 O. S., Page 27
Samuel Jordan is apointed Surveyor of the Nottoway Road in the room of John Nance and it is Ordered that the said Jordan with his Gang clear and keep the same in repair accordingly

10 June 1737 O. S., Page 28
Upon the petition of John Jackson Thomas Jackson Solomon Harper Bryan ffenney William Green William Keatly William Pool and William Parush Leave is given them to clear a Road from the County Line between Tomahitton and the Birchen Swamps to the Chappel on Nottoway the most Convenient way & they are Accordingly Ordered to clear the same.

14 October 1737 O. S., Page 37
William Marshall is apointed Surveyor of the High Waies from Deep Creek to Knibbs Creek & is also Ordered to clear the same with the Gang under him and keep the same in repair.

14 October 1737 O. S., Page 37
Thomas Covington is apointed Surveyor of the High Waies from Tanners to Craddocks & it is Ordered that he with the Gang under him clear & keep the same in repair

14 October 1737 O. S., Page 37
Samuel Hudson is apointed Surveior of the High Waies from Craddocks on Flatt Creek to Bush River & that he with the Gang under him clear and keep the same in repair.

10 December 1737 O. S., Page 41
County Levy
To Thomas Bevil P keeping the fferry ... 520

20 October 1738 O. S., Page 55
County Levy
To Thos.^{s} Bevil ferry keeper ... 800

8 December 1738 O. S., Page 58
At the motion of Joseph Morton It is Ordered that a Road be cleared from George Walkers plantation to Buffalo River and it is Ordered that the people above Bush River together with the s.^{d} Joseph Morton their Surveyor clear the same.

8 December 1738 O. S., Page 58
Ordered that Mr Walkers people clear from the said Walkers plantation into the Road from Coll.^{o} Randolphs Quarters so down as far is before apt.^{d} and that they be not Employed above where the said Walkers road comes into Coll.^{o} Randolphs.

8 December 1738 O. S., Page 58
Mr John Burton is apointed Overseer of the road in the room of Mr John Dawson.

8 December 1738 O. S., Page 58
John Nance jun.^{r} and John Nance Senr John ragsdale Duncomb Hamond and William Hill are Ordered to be Employed on the road over which Arthur Leigh is Surveyor and that they be Exempted from Working on any other road.

8 December 1738 O. S., Page 58
William Mays is apointed Surveyor of the High Waies in the room of Robert Vaughan.

12 January 1738 O. S., Page 61
William Clemment appointed Surveyor of a Road to be cleared from Dabneys to Buckweding the persons on Franklins road and all the other convenient and not employ'd on the other Roads to assist in doing the Same

9 February 1738 O. S., Page 62
Joseph Morton is appointed Surveyor in the room of Samuel Hudson

9 February 1738 O. S., Page 62
William Hudson is appointed Surveyor of West Creek road in room of Wm Clarke

9 March 1738 O. S., Page 63
William Mott is appointed Surveyor in room of William Clarke and leive is given him to turn the road near his house according to the direction of Charles Irby Gent.

20 April 1739 O. S., Page 65
John Leverett appointed Surveyor of the Road from Letbetters low grounds on Nottoway River the nearest way to Butterwood road. Bryan Fenning Wm Jackson Thomas Jackson Hez. Powell John Jackson and all other persons convenient and not employ'd on other roads to assist in doing the Same

20 April 1739 O. S., Page 65
Ordered that the Several Surveyors of publick roads in this County (wherein there Several Roads & Cross Roads meet) forth with cause Posts to be sett up with Inscription thereon in large letters directing the most Noted place to which each road leads

20 April 1739 O. S., Page 66
Joseph Morton Ju^r appointed to mark out and clear a Road from Col.^{o} Richard Randolphs Quarter to the Ridge which Devides this County from Brunswick the nearest and best way all Persons convenient thereto are ordered to assist him in doing the Same

20 April 1739 O. S., Page 66
On the Petition of William Eckhols it is ordered that Paul Pigg Math. Talbert Bat. Austin George Hamm & John Hampton be taken off William Clements Road and ordered on the S.^{d} Eckhols Road

18 May 1739 O. S., Page 67
Richard Jones Gent being appointed by this Court to meet the person appointed by Prince Georg Court to consult and agree with some person for the building a Bridge over Namozain Creek. Reported that they had agreed with Daniel Coleman /for the Sum of Eleven pounds fifteen Shilling/ to build the Said Bridge and that the same be paid after laying the next levey

18 May 1739 O. S., Page 70
The Petition of Sundry Inhabitants and Proprietors of Land in this County for a fferry at the Cut banks is referd until next Court

20 July 1739 O. S., Page 72
James Collins is Appointed Surveyor of the road from Lyles's to Flatt Creek in the room of David Lyles

21 September 1739 O. S., Page 76
Mathew Talbot appointed Surveyor in room of Will. Eckhols and it is ordered that he continue the road from Flatt Creek to the Church and that the same be cleared as soon as conveniently may be and Cosways made where they are wanting

21 September 1739 O. S., Page 76
William Yarbro appointed Surveyor of a Road to be cleared from the head of James Andersons road to the head of Coldwater run upon the Ridge between Nottoway and the Lazaretta the

persons to do the Same Henry Yarbro Christo: Robertson Jn.^{r} Edwd Robertson Rob Rowland and Richard Hix

21 September 1739 O. S., Page 76
Edward Booker Gent is appointed Surveyor of a Road to be cleared the most convenient way from his house to the Church

21 September 1739 O. S., Page 76
John Burton Gent with the gang under him is appointed to continue his road from Flatt Creek to the Courthouse

19 October 1739 O. S., Page 79
Henry Anderson Gent is appointed Surveyor of the road in room of Wm Marshall

16 November 1739 O. S., Page 80
Grand Jury Presentments
... The Surveyor of the road from the Old ponds to William Moles
The Surveyor of the road from West Creek to the Courthouse----

The Surveyor from the Cutt Banks for not keeping the roads in repair --

17 November 1739 O. S., Page 83
Ordered that the road from Bookers Mill be cleared into Lyles's road to the Courthouse.

17 November 1739 O. S., Page 83
John Pride and William Clarke has leive given them to Clear Road for their own use from the said Clarks into the Main road.

17 November 1739 O. S., Page 84
County Levy
Thomas Bevill for keeping a Ferry ... 800
leveyed for Namoz.^{n} bridg & repairing the Boat ... 1400
To John Turner for the directions on the Posts ... 450

21 December 1739 O. S., Page 85
Hugh Boston appointed Surveyor of the Road from the Courthouse into Andersons road in the room of Charles Burk....

15 February 1739 O. S., Page 91
Thomas Forster appointed Surveyor of the road in room of Mathew Talbott

15 February 1739 O. S., Page 92
John & Joseph Jenkins, Nathl Dennis & the hands at John Worshams Quarter ordered on the road over which William Watson Gent is Surveyor.

21 March 1739 O. S., Page 98
Ordered that the road whereon Arthur Leigh is Surveyor be Continued to Nottoway river and that Mr Thomases hands be employ'd thereon. And the road whereon Leverett is Surveyor be Neglected

21 March 1739 O. S., Page 98
On the Petition of Thomas Forster & others praying that the road from the Church to Stocks Creek may be continued to Sandy Creek Richard Booker and John Burton Gent. are desired to View where a Road may be made convenient to the Petitioners and make report thereof

16 May 1740 O. S., Page 102
Ordered that Mr Walkers road be cleared from Saylors Creek to Crawfords and from thence into Burtons road to the Court house

16 May 1740 O. S., Page 102
John Blanchett is appointed Surveyor of the road in room of Wm Shannon ordered that he clear the same according to law

16 May 1740 O. S., Page 102
Grand Jury Presentments:
The Surveyor from the Church to Stocks Creek
The Surveyor from the old pond to West Creek

16 May 1740 O. S., Page 103
On the Petition of Sundry of the Inhabitants of this County between Flatt Creek and Appomatox river praying that a Bridg may be built over the said River where there is a Road already cleared at or near Jenneytoe and the Court being of Opinion that a bridge should be built there accordingly Richard Booker and Joseph Scott Gent are apointed to acquaint the Gent of Goochland County Court with this order ....

16 July 1740 O. S., Page 117
Ordered that William Mayes Surveyor of the Road in the room of Vaughan from the sand Road into Cradocks road the best way he can and that the persons hereafter mentioned assist in doing the Same to (Witt) John Mayes, John Ellis, Jos. Mullord, Edwd Harper James Dicks Fran^s Rice. Robt Vaughan, Hugh Leyton La.^{z} Brown Richd Beisley & William Beisley & that the Same be don as soon as possible

15 August 1740 O. S., Page 119
Charles Irby, Sam¹ Jordan and John Thomas having been appointed by this Court to View where a Road may be Cleared from Brunswick County to Prince George County the nearest and best way and they having don the same do report that begining on Great Nottoway to which Brunswick County has already Cleared a Road from thence along the road already Cleared and then to the deviding line between this and Prince George County according as Thomas Jones hath or shall marke it John Leverett Surveyor & the Same hands employed on his road to do this

15 August 1740 O. S., Page 120
Arthur Leigh is appointed to Clear a Road as Abraham Cock and Samuel Jordan shall direct from Mʳ Cocks Mill into the Church road the same people employed on the Road wher Leigh is before appointed Surveyor are also to do this

15 August 1740 O. S., Page 120
Ordered that a Bridg be built over Wintercomake where the road now is and that the Sheriff give notice that any person willing to under take to build the Sᵈ Bridg do meet at next Court

15 August 1740 O. S., Page 120
Richard Booker Thomas Tabb & Joseph Scott Gent are desired to meet the Gent appointed by Goochland Court to View Appomatox River at Jennitoe and in conjunction with them to agree with an undertaker to build a bridge over the said river at that place

15 August 1740 O. S., Page 120
John Thomas appointed to clear a Road from Jordans Bridg the best way into M.ʳ Cocks road the persons employ'd thereon are Thomas & James Andersons, Thoˢ Taylor & John Thomas all their male Tithables

19 September 1740 O. S., Page 125
William Brown appointed Surveyor of a Road to be cleared from John Martins into Mʳ Walkers road. Martins and Browns hands to do the Same and they are Exemted from working on any other road ...

19 September 1740 O. S., Page 125
Mathew Cabaness appointed Surveyor of a Road to be cleared from James Andersons road Into Jordans road and so to Nottoway Chapell

21 November 1740 O. S., Page 132
George Walker Gent appointed Surveyor of a Road to be cleared from the mouth of Boush river below the mouth of Sandy river into Walkers road. Mʳ Nash Rutledg & Walkers hands to do it

21 November 1740 O. S., Page 132
Grand Jury Presentments
The Surveyor of the road from Smacks Creek to Flatt Creek.
The Surveyor from Stocks Creek to Sandy Creek for not keeping the Same in repair

22 November 1740 O. S., Pages 135-136
County Levy
To John Turner for directions on the posts ... 389

* * *

To Thomas Covington for putting up Posts ... 100

* * *

To Col.º Rich.ᵈ Randolph for building a Bridg ... 1000

* * *

To Thomas Bevill for keeping the Ferry ... 1000

* * *

To Rich.ᵈ Jones for a Bridg over Wintercomake ... 500

* * *

Leveyed for a Bridg &ᶜ for Use of the County ... 3633

20 February 1740 O. S., Page 137
Joseph Morton Sen.ʳ appointed Surveyor of a Road to be cleared from Col.º Richard Randolphs Gent.ⁿ near Hardens the nearest and best way to meet a Road cleared by order Brunswick Court the hands to do the Same are Col.º Randolphs three qᵗʳˢ· Mr Nash's at Camp Creek Hudson Akin and Akin's Tenant

20 February 1740 O. S., Page 138
John Burton Gent being presented by the Grand Jury for not keeping in repair from Stocks creek to Sandy Creek the Said Burton Appearing and not Shewing cause why the Sᵈ road was not repaired It is considered by the Court that he be fined fifteen Shillings and that he pay the Same According to an Act of Assembly in that Case made

20 February 1740 O. S., Page 138
William Barns is Appointed Surveyor of the road from Stocks Creek to Saylors Creek and the Hands above Stocks Creek that was employ'd on Burtons road to assist in doing the Same

20 February 1740 O. S., Page 138
Henry Farloe is Appointed Surveyor of the Road from Stocks Creek to the Courthouse all the hands below Stocks Creek to assist in doing the Same

17 April 1741 O. S., Page 151
Ordered that the Bridle way from Robert Vaughans to the Courthouse be kept open where the path now is

15 May 1741 O. S., Page 157
Richard Hix is appointed Surveyor of Deep Creek Road in the room of Will. Mott and Thomas Jones's hands to work on ye Same

15 May 1741 O. S., Page 157
Robert Moody Surveyor in room of Capt Watson and R^d Beisley to work on that road

15 May 1741 O. S., Page 157
Richard Booker and Thomas Tabb Gent acquainted this Court that the Justices of Goochland County Court refused to Joyne in Agreement with this Court in building a Bridge over Appomatox River at Jennytoe. it is the opinion of this Court that no Suit be brought against the Said Justices for such refusal

16 May 1741 O. S., Page 158
Edward Booker and Saml Cobbs is ordered to agree with some person to remove the Timber allready prepared for building a Bridg over Appomatox River to Websters the place where the Bridge is to be built

21 August 1741 O. S., Page 169
Daniel Coleman appointed Surveyor of the road from Capt Joness quarter to Wintercomake in the room of Robert Coleman

21 August 1741 O. S., Page 169
William Jackson is appointed Surveyor of the road in room of John Leverett

16 October 1741 O. S., Page 175
Joseph Morton appointed Surveyor of a Road to be cleared from John Hudsons to Brunswick County line Mr Nash Collins and Hardens John Cox and John Davis Male tithables to do the same.

16 October 1741 O. S., Page 175
Jacob Magehen appointed Surveyor of the Road to Crafords from John Hudsons. Joseph Mortons Col.^{o} Randolphs hands at Mountain Creek. John Hudson Robert Bowman Thomas Morton and Daniel Browns Male tithables to do the Same

16 October 1741 O. S., Page 175
On the Petition of Sundry Inhabitants above and Below Deep Creek praying that the lower Bridg over the said Creek be repaired whereupon Edward Booker Joseph Scott W^m. Booker and Hezekiah Ford Gtn or any two of them are desired to meet at the place on Saturday the 24th Instant and agree with workmen to repair the Same

16 October 1741 O. S., Page 175
Hands to be aded to those under Benson Surveyor John Hall Jn° Osborn James Atwood Capt. Tabb Stewert & Browns

16 October 1741 O. S., Page 175
Chas. Burk Ju^r. appointed Surveyor in the room of Hugh Boston

16 October 1741 O. S., Page 176
Ordered that Capt. Starke have leave to run the road near his plantation making it as good and convenient as the road that now is

16 October 1741 O. S., Page 176
On the Petition of Saml. Jordan setting forth that the Bridg near Nottoway formerly buil by him being on a publick road and much out of repair may be rebuilt at the County Charge It is ord^d that the same be don and that Charles Irby Gent agree with the builder to do the same

20 November 1741 O. S., Page 177
Henry Anderson Surveyor from Andersons road into Mr Bookers road to the River Bridg the persons under him are Robert Stoker Guy Meak William Hatcher Stephen Neale John Leonard Thomas Leonard John Jackson Richard Walthall and his own people

20 November 1741 O. S., Page 177
Lodwick Tanner Survey of the road to be clear'd from Grangers path in to the road to the Bridg Blanchett & Taylors gangs to assist in opening the road. William Marshall Roger Neale David Neale Branches & S^d Tanners hands to keep it in repair

20 November 1741 O. S., Page 177
Robert Forguson Surveyor of the road from the Harrcane into Jordans road in the room of Jackson the S^d Forguson and Bridgfords hands and all others that hereafter Settle near that road and not employed on other roads to keep the same in repair

20 November 1741 O. S., Page 177
Jacob Seay appointed Surveyor of the road in the room of Thomas Forster

20 November 1741 O. S., Page 177
Thomas Burton appointed Surveyor of the road to Combs's bridg in room of James Collins

20 November 1741 O. S., Page 178
Ordered that Laughland Flyn be excused from working on the roads he being old and infirm

20 November 1741 O. S., Page 178
Grand Jury Presentments
... The road from the Church to West Creek The road from Richard Bookers Mill to Smax Creek the low grounds to Sappony ford ...

20 November 1741 O. S., Page 179
This Court being of Opinion that the lower Bridg over Deep Creek is placed very Inconveniently for the Inhabitants of this County and Daniel Coleman having now repaired it ordered he be paid for the same but the Said Bridg be no more repaired

20 November 1741 O. S., Page 179
Ordered that Charles Irby Gent agree with some person to rebuild the Bridg over the Celler Creek and to keep the same in good repair Seven Years

20 November 1741 O. S., Page 179
On hearing the Petition of William Brown praying leave to clear a road from his house to the Church the best way he can finde ordered he may do the same

20 November 1741 O. S., Page 179
Ordered Thomas Tabb Gent have leave to clear a Road from his House the best way into the road over Appomatox river bridg

20 November 1741 O. S., Page 180
Ordered that John Burton and Hezekiah Ford Gent agree with some person to rebuild Clements Bridg over Flatt Creek

20 November 1741 O. S., Page 180
Ordered William Neale have leave to clear a Road from his house into Andersons road

20 November 1741 O. S., Page 180
William Jackson appointed Surveyor in the room of Her[b] Farley

20 November 1741 O. S., Page 180
Sam[l]. Pincham Surveyor of the road to M[r] Bookers Mill being presented by the Grand Jury and appearing is excused

20 November 1741 O. S., Page 180
John Benson Surveyor of West Creek road being presented by the Grand Jury and appearing is excused

21 November 1741 O. S., Page 181
County Levy
To Thomas Bevill Ferry keeper ... 1000

* * *

To John Turner for lettering boards ... 283

* * *

To William Mayes for puting up four posts ... 50

* * *

To John Benson for puting up bords ... 50

* * *

To pay for Several Bridges ... 7000

18 December 1741 O. S., Page 182
Ordered Henry Walhall Surveyor of a Road to be cleared from Mr Nashes quarter on Boush river into Osborns road. Jno Mullins Charles Cottrell Edward Haskins and Mr Nashes Male Tithables to assist in doing the same

18 December 1741 O. S., Page 182
Ordered a Road be cleared beginning a little below John Winns into Fishers Cart path and from thence to Jordan's bridg Thomas Ellis John Ellis Robert Evans John Evans William Crenshaw and John Mitchell male tithable to do the Same William Evans Appointed Surveyor

15 January 1741 O. S., Page 184
John Hardin is Appointed Surveyor to Brunswick County line in room of Joseph Morton Senr

15 Janury 1741 O. S., Page 185
George Avery Surveyor of the road from Wards quarter into Andersons road Wards Mr Andersons and Wilkinsons hands to the do the Same

15 January 1741 O. S., Page 185
Thomas Covington with the hands under him to Clear the road from Wards quarter to the foot of the hills the other side fflatt Creek and Cradocks Bridg

15 January 1741 O. S., Page 185
Ordered that Jacob Magehee with the people under him clear down Boush river road to where Covington Stops

15 January 1741 O. S., Page 185
Wood Jones & Richard Jones Gent are appointed to agree with some person to build a Bridg over Deep Creek and keep the Same in repair Seven Years

15 January 1741 O. S., Page 186
On the Petition of Sundry of the Inhabitants of this County praying a Bridge may be built over Appomatox River at Capt. Hudson's Quarter where David Lyles lately dwelt The Court on Consideration of the Same are of Opinion that a Bridge over the river there will be very usefull and Convenient to the Said Inhabitants they thereupon do appoint and desire Edward Booker Gent to treat with the Judges of Goochland Court conserning the building a bridg over the same at the most convenient place and that on the Concurrence of the Said Court thereto Mr Edward Booker is further appointed and desired to Joyn with such person as the said Court shall appoint in order that the Same may be build imediately

19 February 1741 O. S., Page 189
Ordered that Joseph Grainger and his Tithables work on the road under John Thomas Surveyor

19 February 1741 O. S., Page 189
Hance Hendrick appointed Surveyor of the Road from Craffords house into Burtons road over Flatt Creek. John Harris William Wilkinson, Wm Brumfeild Wm Farley Jn.^{r} John Hendrick Hance Hendrick Jn^r. & Phillamon Childress to be imploy'd

19 March 1741 O. S., Page 192
Saml. Jordon appointed Surveyor of road from Nottoway Chap^l. to Prince George County line

19 March 1741 O. S., Page 192
John Thomas appointed Surveyor from Jordons Bridg to Cocks road

19 March 1741 O. S., Page 192
Arthur Leigh Surveyor from Nottoway road to the Fork of Nottoway

19 March 1741 O. S., Page 192
William Evans Surveyor from Jordons Bridg to Great Nottoway

19 March 1741 O. S., Page 192
Charles Irby from his House to West Creek

19 March 1741 O. S., Page 192
John Benson from West Creek to the Courthouse

19 March 1741 O. S., Page 192
William Jackson from Great Nottoway to the County line Pr Geo^e and the Church road up to the Harrycain

19 March 1741 O. S., Page 193
Robert Forguson Surveyor from the Harricain to the Chappell

19 March 1741 O. S., Page 193
James Anderson Surveyor from Dandys race paths to Capt Starks new q^{tr}

19 March 1741 O. S., Page 193
William Yarbrow Surveyor up to the Ridg of Nottoway

19 March 1741 O. S., Page 193
George Bagley Surveyor from Spinners to Dandys race paths

19 March 1741 O. S., Page 193
Maj^r. Richard Jones Surveyor from Spiners to Wintercomake

19 March 1741 O. S., Page 193
Daniel Coleman Surveyor from Wintercomake to Namozain Bridg

19 March 1741 O. S., Page 194
On the motion of Richard Booker Gent praying a Road may be cleared from his Mill into the New road to Mr Edward Bookers and the Court believing the same will be a Convenient road do order that Mr Richard Booker with the hands employed on the road over which he is Surveyor clear the road according to the direction of Saml Cobbs

16 April 1742 O. S., Page 197
William Dunifant appointed Surveyor of the Road from Knibbs Creek to the Bridg over Appomatox river the Same hands that were under Saml Pincham to do the Same

16 April 1742 O. S., Page 197
Thomas Brooks appointed Surveyor from Andersons road down to the Bridg over the River. Mr Tanners M^{rs}. Andersons Capt. Worshams M^r. Towns John Jones Will. Belchers male tithables to do the same

16 April 1742 O. S., Page 197
Robert Ferguson appointed Surveyor from Combs Bridg over Flatt Creek to the Court house John Robert & Will Ferguson John Combs & Richard Boram to do the Same

16 April 1742 O. S., Page 197
Saml Cobbs Surveyor of the road from the fork of Burtons road to Knibbs Creek. Jeffersons LeNeves Warrens Edm^d. Bookers and two of Col^o. Harrisons people to do the Same

16 April 1742 Old Style, Page 197
Henry Anderson Surveyor from Knibbs Creek to M^{rs}. Andersons Bridg and to Mr Bookers road Peter & Robert Thompson John Osborn and William Clarke to be added to the gang he before had

16 April 1742 Old Style, Page 197
Ordered that all the Surveyors will Grub & Cut down the Stumps

16 April 1742 Old Style, Page 198
John Burton Gent agreed to keep with his own hands the road in repair from Flatt Creek to the Courthouse for which he is discharged from working on all other roads

28 April 1742 Old Style, Page 199
John Towns presented a petition from Sundry Inhabitants of this County praying a Bridg to be built over Appomatox river at William Towns's plantation ordered the Same be Certified

21 May 1742 Old Style, Page 200
Grand Jury Presentments
... Flatt Creek Bridg at Mr Burtons out of repair
... Knibbs Creek Bridg at Mr Bookers out of repair

18 June 1742 O. S., Page 212
Batho: Austin appointed Surveyor of a Road to be Cleared from Appomatox river near Col^o. Richard Randolphs quarter up to Hills Fork on Vaughns Creek all the persons convenient thereto & not employed on other roads to assist in doing the same

18 June 1742 O. S., Page 212
Henry Dawson appointed Surveyor of a Road to be cleared from William Eckhols's road on Stocks Creek up to the Ridg at the Fork of Sandy Creek John Smith Joel Meadows Richard Loving George Pollard George Foster Paulen Anderson Fran^c Anderson & M^{rs}. Dawsons male tithables to assist in doing the Same

[illegible in book] August 1742 O. S., Page 215
Ordered that Richrd Booker, Abraham Green & William Booker or any two of them agree with some person or persons [illegible] to build a bridge over Deep Creek at Greens

[illegible in book] August 1742 O. S., Page 215
Ordered that a Road be Cleared from the Said Bridge into the main Road the nearest and best Way to Burton's Bridge also Abraham Green Gent is appointed Surveyor of the Same

[illegible in book] August 1742 O. S., Page 216
On the Petition of Sundry Inhabatants of this County praying a Bridg may be built over Appomatox River at or below Capt. Hudsons plantation and the Court believing it necessary a Bridg should be built do appoint Edward Booker Gent to agree with some person for Building the said Bridge at the place afore Said And to desire the Gent. of Goochland Court to Joyn in Agreement with this Court for building the Same and to appoint one or more persons to agree with a builder

17 September 1742 O. S., Page 219
Ordered that the Sheriff pay James Anderson for building a Bridg over Little Nottoway Four pounds fourteen Shillings

17 September 1742, O. S., Page 219
Ordered that the Sheriff pay Edward Robertson for building a Bridg over Cellar Creek four pounds fourteen Shillings and Six pence

17 September 1742 O. S., Page 219
Ordered that the Sheriff pay Hez Ford Gent Four pounds for William Ferguson building a Bridg over Flatt Creek

17 September 1742 O. S., Page 219
Ordered that the Sheriff pay Hance Hendrick for Building a Bridg over Flatt Creek Seven pounds

17 September 1742 O. S., Page 219
Ordered the Sheriff pay Daniel Coleman for repairing Deep Creek Bridg One pound five schillings

17 September 1742 O. S., Page 219
Ordered the Sheriff pay Edward Robertson and Robert Rowland for Building a Bridg over Deep Creek at Peter Jones's quarter Nine pounds

17 September 1742 O. S., Page 219
Francis Anderson is appointed Surveyor of a Road to be cleared begining above and near the mouth of Stocks Creek into the Main Road over Flatt Creek all the persons to whom it is convenient to assist in doing the Same

17 September 1742 O. S., Page 219
Thomas Anderson for his own use has Leave to clear a Road from his House into Thomases Road

15 October 1742 O. S., Page 221
County Levy

To Hance Hendrick for puting up four bords ... 40

* * *

Levy'd for Building Bridges ... 2500

19 November 1742 O. S., Page 222
Ordered that Frederick Ford be Surveyor of a bridle road from Nottoway road to Rocky run Chappell M^r. Bland Christ^o. Hinton Abraham Jones and Maj^r. Munfords Tiths to assist in doing the same But they are not Exempt from working on other roads before ordered

19 November 1742 O. S., Page 222
John Childry Surveyor from Saylors Creek to Sandy Creek all the hands from Sandy Creek Except Dawsons and Thomases that are already ordered on that road to do the Same the Remainer part of the hands to Clear from Sandy Creek to Hurts Creek under W^m. Barnes

17 December 1742 O. S., Page 224
James Rutledge appointed Surveyor of the Road in room of Henry Walters and John Hayes ordered to work on the said road

21 January 1742 O. S., Page 226
On the motion of William Watson Gent It is ordered that a Road be cleared from the head of Little Flatt Creek to the Fork road by Mr Sherwins Plant. Mr Tunstalls hands $Fran^c$ Rice Barnaby Wells Thomas Mitchell $Rich^d$ Beisley John Canister John Willard and Edwd. Harper who is Surveyor and to see that the Same be don

21 January 1742 O. S., Page 226
Joseph Mutloe [Mutlee?] appointed Surveyor of a Road to be Clear'd from the Fork road to Boush river road near Cheathams and the Persons to do it are Hugh Leyton James Dix Saml Jones's $quart^r$. Wm Cradock Edward Covington & Said Mutloes [Mutlees?].

21 January 1742 O. S., Page 226
Christopher Hinton appointed Surveyor of a Road from the Said Hintons into the main Road below Rocky Run Chapell Maj^r. Munfords Abr^a Jones's Abraham Hawks Josiah Hawks $Benj^a$. Bennitts hands to do the Same

21 January 1742 O. S., Page 226
Henry Liggon appointed Surveyor of a Road in room of Geo^{ge}. Walker Gent

21 January 1742 O. S., Page 227
This Court being informed tht One of the Braces of the Bridg built by John Ferguson over Appomatox River was broken and that the Bridg was in Danger of going a Way the Said Ferguson appearing and being asked if he would mend the Said Brace or put it another He refused to do either

18 February 1742 O. S., Page 228
Ordered that $Rich^d$. Booker Gent with hands under him Clear the Road from his mill to the fork of the Road leading to the River Bridg

18 March 1742 O. S., Page 230
Richard Booker and Saml Tarry Gent appointed to agree with a Workman to build the lower Bridg over Flatt Creek

18 March 1742 O. S., Page 230
Thomas Markham appointed Surveyor from Lyles's Ford in room of Tho Burton

18 March 1742 O. S., Page 230
Ordered the Several Surveyors of Roads be continued as they have been Appointed

18 March 1742 O. S., Page 230
Ordered that Edward Booker and Saml. Cobbs Gent bring Suit against John Fergerson for not Building Bridg over Appomatox River according to the agreement made for building the Same

20 May 1743 O. S., Page 232
Grand Jury Presentments
... The Surveyor of the road from Nottoway to West Creek
The Surveyor from Boush River to Saylers Creek by Information of Thomas Osborn
The Surveyor from West Creek to Buckskin a Bord wanting
The Surveyor where Watsons road coms into Flatt Creek for not puting up bords
The Surveyor from Flatt Creek to Appomatox out of repair at the Crossing of West Creek
The Surveyor where Jacksons road coms into the Church road at Nottoway no bords
The Surveyor from Yarbrows to Woody Creek Crossing Irbys road no bords
The Surveyor at the fork of Spinners no bords
The Surveyor at the fork of Cocks and Irbys the same
The Surveyor below Burtons Bridg the Same

17 June 1743 O. S., Page 235
Abraham Cock Gent appointed Surveyor of Road from John Thomas's to the Bridg from Nottoway and to Ridg path near the race path and has agreed it shall go through his Feild by his Peach Orchard the Person to clear the are Henry & William Batts Edward Cox Laugh. Flyn William Cross & s^d. Cox's male Tithables

17 June 1743 O. S., Page 235
Arthur Leigh continue Surveyor from Thomas's to Main Nottoway the Same hands as before excepting those on Cocks and others Convenient

19 August 1743 O. S., Page 246
On hearing the Petition of Sundry Inhabitants of this County for a Bridge between Fighting Creek and William Basses quarter Ordered Joseph Scott William Archer Thomas Tabb and Richard Booker Gent View the most convenient place for building a Bridge and report the Same to the next Court ----

16 September 1743 O. S., Page 247
The Gent appointed last Court to View the most convenient place for a Bridge to be built over Appomatox River above Flatt Creek made report that from the upper part of William Basses Land near Jennytoe in this County to George Williamsons Land in Goochland County would be

a Convenient Place for building the Said Bridge the Court being of opinion that a Bridge over Appomatox River from the Land of William Bass in this County to the Land of George Williamson in Goochland County will be very Necessary and Convenient for the passing from one County to the another and for the Transportation of Tobaccoes to the most convenient Inspections Therefore do appoint Richard Booker Thomas Tabb and Joseph Scott Gent or any two of them treat with the Court of Goochland County to know whether they will joyn in an Agreement with the Justices of this Court concerning the Building a Bridge at the afore Said or whether they will Levy their proportion of the Charge thereof in their County levy according to the Number of Tithables in that County

16 September 1743 O. S., Page 247
Ordered that a Road be cleared from the Ridg road to the place where ti's beleiv'd the Bridge over Appomatox River at Basses will be Built all the hands below the path which goes from Thos Burtons to Lilis's ford except Mich^l. McDearman and Edwd Osborn's be Imploy'd in clearing the Same Joseph Scott Gent Surveyor

21 October 1743 O. S., Page 248
John Nash Gent appointed Surveyor of a Road from Boush river Bridge across Saylors Creek into Walkers Road the Persons to do it are Mr Nash James Rutlidg Rich^d. Rutlidg John Mullins Edmund Gross Charles Cotteril Dougl^s Puckett M^{rs}. Cobb and Edward Haskins's male tithables

21 October 1743 O. S., Page 248
William Marshall Surveyor of the Road from the Lawyers path into Andersons Road the Persons employ'd thereon are Stephen Neale David Neale Roger Neal Josiah Tatum Robert Thomson Peter Thomson John Osborn Witt Clarke Wm Hatchett John Pride Will Marshall & Hugh Chambers

21 October 1743 O. S., Page 248
Joseph Motley with his gang is ordered to Clear a Bridle way out of Boush river road above Mr Reads into Nottoway road

21 October 1743 O. S., Page 248
Anthony Griffin Surveyor of the Road from Watsons muster feild along the Ridge to the first branch of Snales Creek the Person employ'd thereon are Sill Johnson Henry Johnson, Daniel Dejarnett and Capt. Watsons

21 October 1743 O. S., Page 248
Abraham Green Gent Surveyor from the Fork of Bookers Road to the lower Bridge over Appomatox the persons Employ'd thereon are Francis Mann Saml. Morgan Thomas Reams Robert Mann Saml Mann Jno. Perdue Abraham Burton Essex Bevill Thos Bevill Danl Bevill Mr Botts Ralph Perkerson Saml. Pitchford they are also to clear a Road from Deep Creek Bridge to the River Bridge as near Burtons & Bevills lines as possible it can

21 October 1743 O. S., Page 249
John Blanchett appointed Surveyor of the Road from Booker fork to the fork of the Road leading to the uper River Bridge Judah Israel Thomas Brooks Mathew Jackson Danl. Willison Thomas Bottom Lawrence Farley John Willson John Thomson Thomas & Wm. Walkers and William Callicott to work thereon

18 November 1743 O. S., Page 251
The Gentleman appointed by this Court to treat with the Gent of Goochland County Court to Joyn with this County in building a Bridge over Appom^x. River from the Land of William Bass to George Williamsons produced and order from that Court wherein they do refuse to Joyn with this County in building the Said Bridge It is therefore the oppinion of this Court that Suit be brought against the Justices of Goochland County Court for such refusal according to an Act of Assembly in that case made And that Thomas Tabb and Joseph Scott Gent prosecute the Same.

19 November 1743 O. S., Page 253
County Levy
John Turner for Lettering bords Sett up at the Roads ... 959

* * *

Rober Taylor for puting up four bords at ye Roads ... 50

16 December 1743 O. S., Page 255
Lewis Vaughan appointed Surveyor from West Creek into Bush river road in room of William Mayes the persons under him are Robert Vaughan Jeremiah Childry Will Baldwin j^r. Edmund Covington William Mayes and John Ellis all their male tithables

16 December 1743 O. S., Page 259
Ordered all the Male Labouring Persons not employ'd on any Rode Near John Blanchetts be aded to those under him

17 February 1743 O. S., Page 261
Edward Booker Gent has leave to turn the road near his House into his Clear'd ground

16 March 1743 O. S., Page 263
On the Petition of George Moore leave is given him to turn the road near his House provided he clear the Same in such maner as the road now is

20 April 1744 O. S., Page 265
Ordered that Robert Rowland have a bridle way out of Anthony Griffins road to Sandy river Chappell

20 April 1744 O. S., Page 265
Ordered that Joseph Motley with the hands under him clear a bridle way out of Bush river road into Irbys road to the Courthouse

18 May 1744 O. S., Page 266
Mathew Talbott and Clement Read Gent haveing produced an order of Brunswick County Court Impowering them to treat with this Court about building a Bridge over Nottoway River. Ordered that Charles Irby and Abraham Cock Gent. in conjunction with Clement Read and Mathew Talbott Gent do agree with some person to build a Bridge over the Said river persuant to the order of Brunswick County Court

18 May 1744 O. S., Page 267
Grand Jury Presentment
... The Overseer of the Road from Flatt Creek Bridge to Liless ford.

15 June 1744 O. S., Page 271
Ordered that M^r. John Hall have Liberty to Clear a Road from Deep Creek into the Road near James Andersons.

15 June 1744 O. S., Page 271
Edward Thweat is Appointed Surveyor of the Road to be cleared beginning near James Andersons and so into Butterwood Road at or near Leith Creek. William Covington James Hudson & his two Sons William Stanley Bellington Williams Alexander Gray & William Hulm are ordered to clear the same and be excused from Working on other Roads.

15 June 1744 O. S., Page 271
Maj^r. Richard Jones is Appointed Surveyor of the Road over the Head of his Mill in the room of George Bayley and that Bayleys Gang (except Bayley) do clear the same.

20 July 1744 O. S., Page 276
Thomas Markham Surveyor of the Road from Lyles's fford to fflatt Creek being presented by the Grand jury for not keeping the said Road in Repair now Appears and his defence being heard it is the Opinion of the Court that the Presentment be Dismist and that the said Markham do pay the Costs of this Prosecution.

17 August 1744 O. S., Page 281
On Consideration of the Petition of Sundry of the Inhabitants of this County praying that the Bridge over Appomatox River at Abraham Burtons may be repaired the Court being of Opinion that the said Bridge be Repaired accordingly it is thereupon Ordered that Abraham Green Thomas Tabb and William Booker Gent or any two of them do treat with the Court of Henrico about Repairing the same.

17 August 1744 O. S., Page 281
William Watson Gent is Appointed to agree with Workmen to repair the Upper Bridge over fflat Creek.

16 November 1744 O. S., Page 285
On the motion of William Tinstall for a Road to be cleared from his House into the Road leading to Appomatox it is Ordered that the said Tinstall do give notice of the same to the Person or Persons through whose Land the said Road will pass and that they appear at the next Court and make their objections why the said Road should not be cleared.

16 November 1744 O. S., Page 285
Thomas Tabb William Booker and Abraham Green Gent. or any two of them in Conjunction with the Gent. Appointed by the Court of Henrico County are Appointed to agree with a Person or Persons to Build or Rebuild a Bridge over Appomatox River at Abraham Burtons.

16 November 1744 O. S., Page 285
Thomas Bevill is Appointed Surveyor of the Road in the room of Abraham Green.

16 November 1744 O. S., Page 285
On the motion of ffrancis Anderson Praying that a Road may be Cleared from [blank in book] to Clements Mill William Clements and Hezekiah fford Gent. are Appointed to View the Same and make Report to the next Court.

16 November 1744 O. S., Page 285
John Gillintine is Appointed Surveyor of the Road in the room of Jacob Seay.

19 November 1744 O. S., Page 291
County Levy
To John Roberts for going to Bridge to put in a brace ... £ -..3..-

* * *

To John Turner for Letering nine boards ... 214

* * *

To Henry Isbell for Repairing fflatt Creek Bridge ... 200

* * *

To William Dunafant for puting up 4 boards ... 25

* * *

To John Burton Gent. for D° ...4..D° ... 25

* * *

ffor the Bridges and to Defray the money Debts ... 5121

21 December 1744 O. S., Page 292
On the Complaint of John Nash Gent. that the bridge over Boush River is so much out of repair that no wheel Carriage can Pass over the Same with safety. It is thereupon Ordered that Edward Robertson the builder be Sumoned to appear at the next Court to answer the above Complaint.

19 January 1744 O. S., Page 298
On the motion of Edward Booker Junr. leave is granted him to turn the Road through his Land near Nottoway.

15 February 1744 O. S., Page 300
M^r. Richard Clark is Appointed to Inspect the Timbers of the Bridge over Appomatox River and to report his Opinion thereon to the next Court, and that the Sherif do give him notice of this Order.

15 February 1744 O. S., Page 304
Abraham Green and William Booker Gent. being formerly Appointed by this Court to agree in Conjunction with the Gent. Appointed by Henrico Court for the Building a Bridge over Appomatox River now make their Report that they have In conjunction as aforesaid agreed with Thomas Anderson to build the said Bridge and keep same in repair Seven Years for the sum of Seventy Pounds Current Money which said Agreement this Court Confirms.

15 March 1744 O. S., Page 310
Thomas Tabb and William Booker Gent. are Appointed to view the ground below Maj^r. Richard Bookers Mill and report their Opinion to the next Court if a road for Carts may be made there.

15 March 1744 O. S., Page 310
All Surveyors of Roads in this County are continued.

19 April 1745 O. S., Page 311
William Hulm is Appointed Surveyor of the Road in the room of Edward Thweat and Ordered that James Hall Edward Thweat John Grainger and their Tithes do work on the said road.

On hearing the Petition of Henry Robinson for keeping a Cart way from his house to his Mill; Charles Irby Gent. is Appointed to view the same and report to the next Court what Damage it will be to any Person or Persons by keeping the said road.

On the motion of William Watson Gent. it is Ordered that a Bridle way be opened from the New road at the head of Lazaretto to Nottoway Chapple and that the said Watson, Mathew Cabanis, James Olive, Edward Robertson, Nathaniel Robertson & Isham Vaughan and their Tiths do open and clear the same & that Matthew Cabanis be Surveyor thereof

James Olive is Appointed Surveyor of the New road from Mallerys Creek to the Race paths below Watsons -- and that Peter Jones's Tenant, Watson's and all the Tiths convenient to the said road clear the same.

Robert Vaughan is Appointed Surveyor of the ffork Road from Irby's Court house road into Boush River Road and that John Ellis, William Mayes, Widow Childry, Lewis Vaughan Mathew Mayes, John Childry & their Tiths do clear the same.

Ordered that Richard Hix and the hands under him do clear the Road from where Irby's Road Crosses the Road to Mayes's down to Deep Creek. and that Thomas Bottom do work on the said Road.

Ordered that a Road be Clear'd from the Rev^d. M^r. John Ornsby's to Dandys Race Paths. William Dandy William Shorts, Thomas Williams, Ephrim West, John West, John Clark, William Hardcastle and their Tiths are to clear the said Road and that William Dandy be Surveyor thereof.

John Mayes is Appointed Surveyor of the Road from Dandys Race Paths into Thomas Jones's Road and that William Westbrook, Charles Westbrook, Curtis Cates, Richard Dennis, David Ellington and their Tiths, and the Tiths at Halls, Wards and Hudsons Quarters do clear the Same.

Ordered that a Road be Cleared from Mallarys Creek along the Ridge to Randolphs Road at the head of Boush & Maherrin Rivers and that Anthony Griffin & the hands under him, Stephen Collins Peter Davis, John Hayes and their Tiths & the Tiths at Col^o. Richard Randolphs Quarter where Harding is Overseer and at Capt. John Nash's Quarter do open and clear the same.

19 April 1745 O. S., Page 311
On the Petition of Richard Booker Gent. setting forth that a better & more convenient Road than that Road that now leads over his Mill dam may be had, and that he will at his own Expence clear and maintain the said Road, and build a Bridge over the said Millstream sufficient for Wheel Carriage & keep the same in good repair. The Court having formerly Appointed two of their Members to view the Ground for the said Road now make their report, that a new Road for the good and sufficient Bridge over the said Mill Stream will be as convenient as the Road that now leads over the said Mill Dam -- Whereupon it is Ordered and leave is granted the said Richard Booker to open and Clear a new Road below his Mill Dam, and that he build a good and sufficient Bridge over his Mill Stream at his own Expense at the Said Road Sufficient for Carts and other Wheel Carriages to go over and keep the same in repair and it is also Ordered that the old Road Leading over the said Mill Dam be Stopped.

Edmund Booker is Appointed Surveyor of the Road in the room of Richard Booker Gent. and the said Richard on his building a Bridge over his Mill stream and keeping the same with the new Road in repair his Titheables at the Plantation whereon he now lives are exempted from working on any other Roads.

17 May 1745 O. S., Page 312
On the Petition of Henry Robertson praying that leave may be granted him to Clear a Cart Path from his House to his Mill, Charles Irby Gent. having viewed the Ground through which the said Path is to be cleared in Pursuance of the last Courts order and no Person appearing to make any objection thereto it is considered that leave be granted him to clear the said Path and that Charles Irby Gent. direct the same.

17 May 1745 O. S., Page 313
Grand Jury Presentments
... the Surveiors of the Roads from Stocks Creek to fflat Creek, from Cocks Quarter to West Creek, from Bookers Mill to the ffork of the Road to the River Bridge, from James Cheathams to Judis Branch, George Booker for turning the ridge Road and is out of repair. Ordered that the Several Surveiors (except Booker and Gillingtine) be Sumoned to appear at the next Court to Answer the Presentments.

17 May 1745 O. S., Page 314
Ordered that a Road be cleared from the Head of Little Ronoak along the Ridge between Briery and Buffalloe Rivers to Rutlidges fford over Appomattox River. Charles Rickey, Hugh Rickey, Alexander Rickey, ffrancis Rice, Richard Woodson, Ralph Elkin, John Hudson, John Bibb, Joseph Morton, John Martin, Hugh Nixon, John Mullins, Charles Anderson and their Tiths and all other Persons convenient thereto and not employed on other Roads are to clear the same And that Charles Anderson be Surveyor thereof.

17 May 1745 O. S., Page 315
Ordered that Benjamin Hendrick William Edwards William Silcock, Henry Nelson, Thomas Nelson, George Moor, William Jones, John Harris, William Russell, William ffarley Junr., James Broomfeild Junr. John Hendrick and James Broomfeild be added to the Gang of which Hance Hendrick is Surveyor

18 May 1745 O. S., Page 316
Ordered that a Road be Cleared out of Bush River Road a litle below John Braggs to the Church and that John Davis be Overseer thereof And that Stephen Collins, Richard Womack, Col^o. Richard Randolphs Tiths at Mountain Creek Quarter, George Steward, John Braggs, John Hudson, James Ball, and their Tiths Assist in Clearing the same.

Ordered that Jacob Mcgehee make such Divisions in the Road whereof he is Appointed Surveior And Apportion the Hands to each Division as he shall think fit.

21 June 1745 O. S., Page 321
It is Considered that all such Male Labouring Tithable Persons as lives in this County and within two Miles of Appomatox River and are willing to Assist in Clearing the said River be Exempted from clearing and repairing the High Waies and that the Owners of such Mills as are already Erected or are now about to be Erected on the said River keep open such Passages thro' their Mill Dams as is convenient and Necessary for the safe passing of such Boat or other Vessell as shall go thro' the Same.

Edmund Gross is Appointed Surveyor of Appomatox River from the Mouth of Buffoloe River to the Mouth of Bush River on Appomatox River.

Duglas Pucket is Appointed Surveior of Appomatox River from the Mouth of Bush River to Townes Quarter on Appomatox River.

Henry Dawson is Appointed Surveior of Appomatox River from Townes's Quarter to Lovells Mill on the sd. River

John Echols is Appointed Surveior of Appomatox River from Lovell's Mill to Clements's Mill on the sd. River

Thomas Lorton is Appointed Surveior of Appomatox River from Clements Mill to Little Let alone on the sd. River

John Mann is Appointed Surveior of Appomatox River from Little Let alone to the Mouth of ffighting Creek on the said River

William Townes is Appointed Surveior of Appomatox River from the Mouth of ffighting Creek to the Mouth of fflat Creek on the s^d. River.

Stephen Crump is Appointed Surveior of Appomatox River from the Mouth fflat Creek to the uper Bridge on the said River.

Zecheriah Bell is Appointed Surveior of Appomatox River from the uper Bridge to the Lower Bride on the s^d. River

Abraham Burton is Appointed Surveyor of Appomatox River from the Lower Bridge to Wintocomacke Creek on the said River

George Tucker is Appointed Surveior of Appomatox River from Wintocomacke Creek to Namoseen Creek on the said River

22 June 1745 O. S., Page 324
William Jackson is Appointed Surveyor of the Road from Hurts Creek to fflat Creek.

22 June 1745 O. S., Page 324
The Presentment of the Grand jury ag^t. William Dunafant for not Clearing a Road to Appomatox Bridge is Continued.

22 June 1745 O. S., Page 325
On the Petition of George Booker leave is granted him to turn the Road near his House on Condition that he make the new Road as good as the old.

19 July 1745 O. S., Page 328
Ordered that a Road be Cleared beginning against Edward Jones's and to Cross West's Creek near Tallys Branch and from thence between Wilkinsons Quarter and Wards the best and nearest way to the Court House. Edward Jones, John Osborn, Stephen Bentley and their Tiths, both the Wilkinsons, Wards and M^r. Hardaway's Tiths are to Clear the Same and that Mr. Hardaway be Surveior thereof.

17 August 1745 O. S., Page 337
Thomas Anderson informs the Court that he had Built & finished the Bridge over Appomatox River at Burtons by him undertaken to build the said Thomas Anderson together with Samuel Tarry Clement Read and George Currie his Securities enter into Bond for his keeping and maintaining the said Bridge in good repair for and during the term of Seven Years Pursuant to an Agreement made by the said Thomas Anderson with Persons Appointed by this Court and the Court of Henrico and now moves that the said Bridge may be received It is thereupon Ordered that Thomas Tabb Abraham Green and William Booker Gent. do View the said Bridge and if it Appears to them to be well and sufficiently built that then they receive the same on behalf of this County And it is also Ordered that on receiving the s^d. Bridge that Abraham Green Gent. late Sheriff do pay unto the said Thomas Anderson out the Money of Tobacco of this Counties in his hands the Sum of fffifty pounds Current Money being this Counties Proportionable part for building the said Bridge.

18 October 1745 O. S., Page 343
Ordered that a Road be Cleared from Bush River Bridge to the Chapple M^r. Nash's M^r. Walker's & M^{rs}. Cobbs Tiths Henry Ligon William Ligon Alexander ffrazier, James Moor James Rutlidge Charles Cottrell & their Tiths are to Clear the same And Douglas Puckett is Appointed Surveyor thereof.

Ordered that William Barnes with hands under him do Clear a Road from or near Whitworths in Saylors Creek Road below the Race paths and that M^r. Merediths Tiths and Thos. Whitworth Junr. be added to the said Barnes's Gang.

On the petition of William Yarbrough leave is granted him to Clear a Road from his House into Capt. Irby Road to the Court House.

John Talley is Appointed Surveyor of the Road from Namozain Bridge to James Coles Spring Branch and Robert Tucker is Appointed Surveyor of the Road from James Coles's Spring Branch to Wintercomacke the hands that belong to Tucker's Gang are to be kept divided between them.

Daniel Coleman is Appointed Surveyor of the Road from Wintercomacke to Rockey Run.

Thomas Booth Junr is Appointed Surveyor of the Road from Rockey Run to Spencers branch and Ordered that the hands which were under Coleman be equally Divided between Booth & the said Coleman having regard to what is wanting to be done on each Road.

Ordered that John Prides hands do work on M^r. Henry Andersons Road.

18 October 1745 O. S., Page 343
On the motion of Abraham Green Gent. Leave is granted him to Clear a Bridle way from his House to Rockey Run Chapple.

15 November 1745 O. S., Page 344
Grand Jury Presentments
... the Road from fflatt Creek Church to M^r. Henry Andersons is out of Repair and we also present the Surveyor thereof for the same

... the Road that Robert Vaughn is Surveyor of is out of repair and we also present him for the same

20 December 1745 O. S., Page 349
Henry Anderson being Presented by the Grand jury for not keeping the Road in repair Appears and his excuse being heard he is thereupon Acquitted.

20 December 1745 O. S., Page 349
Robert Vaughan being Presented by the Grand jury for not keeping the Road in repair Appears and his excuse being heard he is thereupon Acquitted.

17 January 1745 O. S., Page 352
Ordered that a Road be Cleared out of the New Road near Mallarys Creek in to Sandy River Road and from thence to the Church and that Daniel Brown be Surveyor thereof Charles Burks Joseph Ligon Thomas Morton William Seircey James Gravely Matthew Womack and their Tiths are to Clear the same.

Ordered that Josiah Hudson Osborn Keeling and ffrancis Spellers & their Tiths be added to Mathew Cabanis's Gang on the Church Road.

Stephen Collins is Appointed Surveyor of the Upper part of the Road of which [blank in book] Griffin is Surveyor and that the Hands that work on the said Road be Divided between them

21 March 1745 O. S., Page 359
The Several Surveyors of the Roads are Continued.

21 March 1745 O. S., Page 359
Ordered that Samuel Tarry Gent. Sherif do imploy Workmen to repair the Bridge on fflat Creek near William Craddocks Nibbs Creek Bridge and Combs Bridge on fflat Creek and that the charge of such repair be amounted for at the Laying of the next County Levy.

21 March 1745 O. S., Page 360
Information being made to this Court that the Bridge over Appomatox River is so much out of repair that it is Dangerous passing over with Waggons and other Wheel Carriages. Whereupon Thomas Tabb and Joseph Scott Gent. are Appointed to treat with the Court of Henrico about rebuilding or Repairing the same and to agree in Conjunction with the Persons appointed by the said Court With Workmen to Rebuild the said Bridge.

18 April 1746 O. S., Page 361
Ordered that a Road be Cleared from Stocks Creek to Sandy Creek and that John Maulden be Surveyor thereof. John Smith Sen[r]. & Jun[r]. Richard Loving Thomas ffoster John fforster Joseph Seay Thomas Green and their Tiths are Ordered to Clear the same.

18 April 1746 O. S., Page 362
Ordered that Thomas Bevil Surveyor with the hands under him do clear the Road from his fork to Old Saponey fford and make a Bridge to the same and cut down the Banks at the River sufficient for Carts and other Wheel Carriages to pass with safety.

18 April 1746 O. S., Page 362
Ordered that Samuel Tarry Gent. Sherif do Agree with some Person to Rebuild the upper Bridge over fflat Creek Immeadiatly.

Amelia County Order Book II

16 May 1746 Old Style, Page 1
Robert Baker Surveyor of a Road to be Clear'd from Sandy fford on Appamatox River to the main branch of Spring Creek the Persons to do the Same are Colo. Richard Randolph's Col[o]. Will[m]. Randolph's James Cunningham John Cunningham Sam[l]. Evan James Alexander James Ewen Sam[l]. Baker John Thompson James Parks Joseph Little John Robert Martin and [page torn] Persons convenient thereto & not employ'd on any other Roads .....

16 May 1746 Old Style, Page 1
Peter Wynne Appointed Surveyor of the Road from the Bridge over Nottoway River the best way to the Chapel Road the Persons to do the Same are James Crowder, William Pitman ju[n]. John Pitman one of Edward Cox and all Thomas Bowry's

16 May 1746 Old Style, Page 1
Henry Ward Apointed Surveyor of the Road in the Room of John Benson

16 May 1746 Old Style, Page 1
Ordered that Maj[r]. Blands Tiths be taken off the road to Namozain Church and to Assist in Clearing the Road from the Said Blands Quarter to Dandys Race Paths John Stegall Surveyor of the Same .....

16 May 1746 Old Style, Page 1
Ordered that Capt. Watsons hands be aded to ffred^k. ffords Gang .....

17 May 1746 Old Style, Page 1
Abram Green and Wood Jones Gent. are Apointed to Treat with the Persons Apointed by the Court of Prince George to agree with Some Person to rebuild Namozein Bridge

16 May 1746 Old Style, Page 2
Grand Jury Presentments
... William Clement for Stoping Appamattox River by a Mill. John Smith for Stoping Appamattox River by a Mill. The Keeper of the upper Bridge over fflatt Creek ---

17 May 1746 Old Style, Page 7
Thomas Tabb Joseph Scott and William Booker Gent or any two of them in Conjunction with the Gent Appointed by the Court of Henrico County are Appointed to agree with a Person or Persons or Persons to Rebuild the Bridge over Appamattox River at Goodes.

20 June 1746 Old Style, Page 8
Edward Booker Gent. is Appointed to agree with some Person to rebuild the upper Bridge over fflatt Creek.

15 August 1746 Old Style, Page 19
ffrancis Anderson is Apointed Surveyor of the Road from Gillintines to Stocks Creek and from Clements Old Mill to Gillintines the following Persons are Apointed to Assist in doing the Same (to wit) Paulin Anderson Ben. Clements Jesse Seay Jno. Hurt William Eckhols, Richard Anderson, William Clement Junr. Thomas Hardy and the Said ffran^s. Anderson and all their Male Tithables et^c.

20 September 1746 Old Style, Page 21
Ordered that Charles Burks Senr. John Cole and all their Male Tithables do assist in Clearing the road over which Daniel Brown is Surveyor

20 September 1746 Old Style, Page 22
Ordered that Thomas Anderson be Sumoned to appear at the next Court to Answer a Complaint made ag^t. him about the Lower Bridge over Appamattox River

20 September 1746 Old Style, Page 23
Ordered that M^r. Clements & M^r. fford do agree with Some Person to repair fflatt Creek Bridge

17 October 1746 Old Style, Page 26
Ordered that Thomas Hudson and all his Male Tithables be employed on the road from fflatt Creek to the Courthouse

24 November 1746 Old Style, Pages 28-29
County Levy
To Thomas Webster for taking care of the Bridge over Appamx River   £2..---..---

* * *

To Hance Hendrick for Building a Bridge over fflatt Creek   £4..---..---
To John ffarguson for repairing a Bridge over fflatt Creek   £2..16..11.

16 January 1746 Old Style, Page 30
On the motion of John Booker Praying Leave to Stop the road near his house that is the way to Richard Booker's Mill and that the Old road near John Comb's be kept open and that he the said John Undertakes to make a Bridge over the run near the said Combs's and Constantly keep the same in Good repair. In Consideration whereof Henry Anderson ffrancis Anderson & Lodwick Tanner are apointed to View both roads and report their Opinion thereupon which is to be the Judgment of this Court.

16 January 1746 Old Style, Page 30
George Moore is Appointed Surveyor of a road from Snails Creek to Lunenburg line in the room of Stephen Collins and that John Hardin George Stewart Stephen Collins M^r. Nashes Quarter Col^o. Randolphs two quarters Robert Atkins Watkins's Davis's and all other Persons convenient thereto and not Employed on any other road are Ordered to clear the same

16 January 1746 O. S. Page 30
George Evans Apointed Surveyor of the road from Stocks Creek to the Church in the room of John Gillington

16 January 1746 O. S., Page 30
John Ellington Apointed Surveyor of a road to be Clear'd from Hudsons on West Creek down into Manns road and that Stephen Beasly Robert Thompson William Clarke M^r. Prides M^r. Branches Geo. Stegall Hugh Chambers M^{rs}. Crawley's William Osborn Peter Rowlett Robert ffarguson Peter Irby Robert ffield and all their Male Tiths do Assist in Clearing the same

16 January 1746 O. S., Page 30
Ordered that Joseph Scott Gen^t. apply to the Court of Goochland County to know if they will Join in an Agreement to build a Bridge over Appamattox River from the Land of Willm Bass in this County to Williamsons Land in Goochland County.

20 February 1746 O. S., Page 31
George Booker is Apointed Surveyor of the road from Buckweding in the room of William Clement Gent.

20 February 1746 O. S., Page 31
Joseph Scott and Thomas Tabb Gent. in Conjunction with the Gent. Apointed by Goochland County Court are Appointed to agree with an Undertaker to Build a Bridge at or near Genetoe and Warrant the Same if they think Proper.

20 February 1746 O. S., Page 31
William Hutcherson is Apointed Surveyor of the road in the room of Thomas Markam

20 February 1746 O. S., Page 31
On the motion of Edward Booker Gent. leave is given him to turn the road round M^r. Smith's fence on the said Smith's Land.

20 February 1746 O. S., Page 32
Robert Cousens is appointed Surveyor from Namozain Road to a Chestnut Oak on William Greens Land All the Persons below Wintercomake and under William Green to assist in clearing the Same

20 February 1746 O. S., Page 32
Edward Reams is appointed Surveyor from the Chestnut Oak to Deep Creek Bridge all the Persons above Wintercomake under William Green and Abraham Greens hands to do the same

20 February 1746 O. S., Page 32
Joseph Morton Ju^r. is Appointed Surveyor of a Road to be cleard from Hudsons ford on Buffilloe up the Ridge Opposit to Col^o. W^m. Randolphs uper quarter and from the same ford into the Road to Rutledges ford near the School house. Edwd. Brathwett Saml. Baker Saml. Wallis George Caldwell Robt. Galespy George Ewing James Alexander Rober Cambell and Col^o Will. Randolphs tiths to clear the Same

20 February 1746 O. S. Page 32
Richard Woodson Surveyor of a Road to be clear'd from Bush River Bridge into the Road leading to Rutledges Ford. John Gaulden John Watkins Theo Carter Joseph Shelton Richd Morton M^{rs}. Nash M^{rs}. Cobb and Said Woodsons tiths to do the Same

20 February 1746 O. S., Page 32
Ordered the Sheriff pay Abraham Green Gent. what money is due from this County to Abraham Talley for rebuilding Namozain Bridge

20 March 1746 O. S., Page 35
William Farley is appointed Surveyor of a road in the room of William Jackson.

21 March 1746 O. S., Page 37
Edmund Booker Junr. Appointed Surveyor of the road from Bookers Fork to Flatt Creek Bridge with the same Hands under his Father also to clear a road out of the road to John Bookers to

cross over or near the mouth of his Brick-yard Branch into the main road to Col°. Bookers and that John Booker's Tiths do Assist in Clearing the Same.

21 March 1746 O. S., Page 37
All Surveyors of Roads in this County before apointed are Continued.

21 March 1746 O. S., Page 39
William Watson appointed Surveyor of a Road to be Cleared from Spring Creek to Lunenburg Line the Persons convenient thereto to clear the same.

17 April 1747 O. S., Page 39
Charles Irby and Abraham Cock Gent are appointed to meet the Persons appointed by the Court of Lunenburg to Joyn with them in Building a Bridge over Nottoway River where Wininghams Road croses the Same.

15 May 1747 O. S., Pages 40-41
Grand Jury Presentments
... The Surveyor of the Road from Flatt Creek lower Bridge to Col°. Cobbs ordinary for not keeping the same in repare
The Surveyor of the Road from Goods Bridge to Col°. Bookers
The Several Surveyors of Bush river road
the Surveyor of the Road from Snales Creek to the County line

* * *

The Surveyor of the Road from Knibs Creek to Andrews Bridge
We present the Surveyor from Snales Creek Cross Mallarys Creek
The Surveyor from Deep Creek Bridge to Capt Irbys Cross Road

* * *

The Surveyor of the Road to Cross Flatt Creek upper Bridge

15 May 1747 O. S., Page 41
Thomas Pettis Surveyor from Flatt Creek Bridge to Southalls Ordinary John Combs John Farguson Richd. Boram and S^d Pettiss Roger Thomson to do the Same

15 June 1747 O. S., Page 42
[Trial of Will, a slave of John Hudgins, convicted of murdering Jack, another of Hudgins's slaves. Will was found guilty and sentenced to be hanged; ordered that after his death:] his Head be severed from his Body and fix'd on a Pole at a fork of the Road near Southall's ordinary.

20 June 1747 O. S., Page 45
Abraham Green and Wood Jones Gent are appointed to agree with some Person to Build a Bridge over West Creek near Hudsons Cart way.

21 August 1747 O. S., Page 49
Samuel Sunderland Appointed Surveyor from Sandy Ford to Col°. Randolphs Mill and Vaughans Creek

22 August 1747 O. S., Page 51
Ordered Saml. Tarry Gent the late Sheriff out of the money in his hands pay Thomas Tabb Gent this Countys Proportion of the Charge for Building a Bridge over Appomatox River at Jennytoe

17 September 1747 O. S., Page 53
John Chisham Appointed Surveyor from George Hams into the Road to Saylors Creek Nicholas Gillintine Mathew Hilsman Jn° Hill Jacob Seay George Hamm Saml. Major John Major and John Clemment all their Male Tithables are to clear the Same

16 October 1747 O. S., Page 55
William Belcher is appointed Surveyor of the Road cal'd Tanners Worsham. Towns. Brooks. John Belcher and George Goodwins Male Tithables to clear the Same.

16 October 1747 O. S., Page 55
Robert Thompson appointed Surveyor of the Road from West Creek road Just below Hudsons Race paths down to Hugh Chambers Plantation

16 October 1747 O. S., Page 55
John Ellington Appointed Surveyor from Chambers to Neals race paths.

16 October 1747 O. S., Page 55
Francis Mann Appointed Surveyor from Neals to Sappony Road — the Surveyors are to take the People most convenient to the Several Roads

16 October 1747 O. S., Page 55
James Jackson Appointed Surveyor of a Road to be Clear'd from Nottoway Bridge to Butterwood Spring into Cocks Road. The hands employd on part of that Road and all others convenient thereto and not on other Roads are to Clear the Same

16 October 1747 O. S., Page 55
Thomas Yarbrough Junr. is Appointed Surveyor of a Road to be Clear'd from Henry Robertsons Mill path to Crenshaws ford over Little Nottoway into Jordans road below the Chapell. Thos Yarbrough Hen. Yarbrough John Hayes Phil Pledger Henry Gaines & John Johnsons hands to work thereon

16 October 1747 O. S., Page 55
William Brown Surveyor of a Road to be Clear'd from Bush River Road a Little below the Pole Bridge along the Ridg into Mallorys Creek road. Lester. Siresey. Potter. Tho Morton & Ambros Beisleys hands to clear the Same

16 October 1747 O. S., Page 55
George Foster Surveyor of a Road to be clear'd from Bush River Road at Beisleys Path Crosing the Creek below Watsons Mill thence into Watsons Road below Tunstalls Quarter Richd Beisley John Dyer Jo^s Yarbrough John Roberts Major Willis & William Branton [Branson?] to Clear the Same

16 October 1747 O. S., Page 55
John Towns Appointed Surveyor in room of Joseph Scott Gent.

16 October 1747 O. S., Page 55
James Atwood has leave to Clear a bridle way from Little Bryer River into the Chapell road

13 November 1747 O. S., Page 56
County Levy
To M^r. Tabb what he advanced to pay for Goods Bridge ... 1900

20 November 1747 O. S., Page 57
Ordered that John Taylor be Appointed Surveyor of the Road in the Room of Samuel Gordon who is discharged from that Office ---

20 November 1747 O. S., Page 57
Ordered that Charles Irby Gent do View the Road lately made thro Edward Booker Junior's Land and make Report to the Court what Damages this Said Booker is likely to Sustain by Occasion thereof ---

20 November 1747 O. S., Page 57
Ordered that Anthony Griffins & his Male Labouring Tithables together with George Jones John Owens & John Carrell Assist George Moore in Clearing his Road from Snails Creek into the Said Moore's House & up to Col^o. Richard Randolphs Upper Quarter likewise that the Said Moore keep the Same in Repair ---

20 November 1747 O. S., Page 57
Ordered that Joseph Rice Surveyor the Parson Thomas Turpin John Holloway Richard Witt Michael Rice John Waddall & their Male Labouring Tithables Clear a Road from Sandy River where Capt. Walkers Old Road Cross'd it the best & Nearest Way in to Bush River Road

20 November 1747 O. S., Page 58
Grand Jury Presentments
... George Moore for turning the Road by his House & Runing a fence Over the Old Road Contrary to Law - likewise against the Surveyor of the Road from George Moore's to Snails Creek the said Road being out of Repair

20 November 1747 O. S., Page 60
Ordered that William Booker Gent View the Road leading thro' Benjamin Ward's Plantation which road the Said Ward purposeth to turn Round his Said Plantation and make Report thereof to the Next Court ---

20 November 1747 O. S., Page 60
Ordered that John Combs be Appointed Surveyor of the Road in the Room of Ed^{md} Booker Junior who is discharged from that Office ---

21 November 1747 O. S., Page 61
Ordered that William Womack be Appointed Surveyor of the Road from Great Saylor into the Road a Little below Crawfords House And that Thomas Certain John Nash's Quarter Benjamin Ruffin's Quarter Abraham Vaughan John Gentry John Howell William Brooks & Charles Spradling with their Male Labouring Tithables Work under the Said Surveyor on the Aforesaid Road And that he follow the direction of M^r. Nash as to the Alteration of the Road ---

21 November 1747 O. S., Page 61
Ordered that M^r. Booker & M^r. Cobbs View the Most Convenient Way where the Road may be turned by M^r. Burton's ---

18 December 1747 O. S., Page 65
Ordered that Thomas Tabb & Joseph Scott Gent View the Bridge lately Built by James Anderson and if the Said Bridge is Compleated in a Workman like manner that they Receive it from the Aforesaid Anderson It is further Ordered that After they have Received the Said Bridge that they Agree with Some person or persons to keep the Same in Repair And that they likewise make Application to the Gentlemen of Goochland County to join with them in the Expence of the Said Bridge and make Report thereof to the Next Court ---

18 December 1747 O. S., Page 68
Ordered that Dasey Southall be Appointed Surveyor of the Road in the Room of John Combs who is discharged from that Office ---

15 January 1747 O. S., Page 68
George Moore being presented by the Grand Jury for turning the Road Near his house for Certain Good Causes Appearing to the Court is Excused ---

15 January 1747 O. S., Page 68
On the Petition of W^m. Kennon John Robertson & James Robertson praying that the Road from John Robertsons may be Continued from the Said Robertsons a Cross the Flatt Creek thro' the Land of John Gibbs & Essex Worsham into the Road that Leads to Goods Bridge And it is Ordered that Col^o. Booker Richard Booker & William Booker Gent do View the said Road and make Report thereof to the Next Court ---

15 January 1747 O. S., Page 68
Ordered that Francis Anderson & George Booker Assist George Evans with the hands Under their direction to make a Cause Way at Flatt Creek Bridge as Often as it is Wash'd away ---

15 January 1747 O. S., Page 68
Ordered that Charles Burke be Appointed Surveyor of the Road that is to be Cleared from the Courthouse into Andersons Road and that Thomas Porter George Burke Daniel Lewelling James Mitchell & Cox Quarter Work under the Said Surveyor in Clearing the Aforesaid Road ---

15 January 1747 O. S., Page 68
Ordered that Col°. Booker Agree with Some Person to Rebuild the bridge over Knibbs Creek Near his house ---

19 February 1747 O. S., Page 68
On the Petition of William Jackson & others for the Clearing & opening a Road from Batts Path to this County line / the Most Convenient Way / is Granted the Said Petitioners Upon their Clearing Such New Road as the Law directs ---

19 February 1747 O. S., Page 69
Ordered that Peter Davis & the hands Under him with the hands at Col°. Randolph's Quarter & M^r. Nash's & George Stewarts Assist George Moore in Clearing his Road

19 February 1747 O. S., Page 69
On the Petition of Col°. Booker Ordered that Liberty be Granted him to turn the Road Near his house ---

19 February 1747 O. S., Page 71
On the Petition of Joseph Ward Ordered that he have leave to turn the Road from his house into Sailors Creek Road ---

18 March 1747 O. S., Page 74
Ordered that Daniel Degarnett be Appointed Surveyor of the Road in the Room of Anthony Griffin who is discharged from that Office ---

18 March 1747 O. S., Page 74
Ordered that Flouronoys hands do Assist Peter Davis in Clearing his Road ---

18 March 1747 O. S., Page 74
Ordered that Henry Ligon Duglas Puckett & Thomas Williamson be Appointed to View Bush River Bridge & make Report thereof to the next Court —

18 March 1747 O. S., Page 74
Ordered that the Sherif be Appointed to Agree with Some Person or Persons to Repair the Upper Bridge Near Flatt Creek ---

18 March 1747 O. S., Page 75
Ordered that Wood Jones Gent Agree with Some Person or Persons to Build a Bridge over West Creek on Hudsons Road and that the Said Undertaker or Undertakers Shall Compleat the same in a Workman like Manner.

18 March 1747 O. S., Page 76
Ordered that Stith Hardaway have Liberty to Clear a Road into the Road leading to the Bridge Over West Creek ---

18 March 1747 O. S., Page 76
On the Petition of Henry Clark William Stone William Manire Lewis Hammond Robert Taylor George Hill & Charles Connolly for a Road from the County line to Jacksons Road It is Ordered that M^r. Irby & M^r. Cocke direct what Way is the Most Convenient for the Clearing the Said Road & what hands are to do it ---

18 March 1747 O. S., Page 76
Daniel Coleman personally Appeared in Court & Purposed to keep Wintercomack Bridge in Repair Seven Years from September Last and to leave the same at the Expiration of that term in good repair for the performance of which he is to be paid five hundred pounds of Tobacco ---

18 March 1747 O. S., Page 78
Nicholas Guillington being Summoned to Appear at this Court for turning the Road Upon Mature Deliberation being thereupon had the Said Nicholas is Excused ---

21 March 1747 O. S., Page 82
Ordered that John Booker's Roger Thompson's John Pride's & Field Jefferson's Male Labouring Tithables do Assist Samuel Cobbs Gent with the hands Under him in Clearing his Road ---

21 March 1747 O. S., Page 82
Ordered that William Brown William Watson and George Forster Gent do View the Ridge Road that Anthony Griffin was Overseer of and Make Report to the Next Court

15 April 1748 O. S., Page 84
On the Petition of John Morris It is Ordered that Richard Booker Gent & William Ligon do View the Road Leading from Joseph Wards to Sailors Creek Road & Make Report thereof to the Next Court —

15 April 1748 O. S., Page 84
On the Petition of Charles Anderson & Other the Inhabitants of the Upper Part of this County for a Bridge to be Built over Appamatox River a little above the Mouth of Bush River Opposite to the Land of Col°. Peter Randolph the Court being of Opinion that the Same is reasonable And It is Ordered that Clement Reed Gent be Appointed to treat with the Gent of Goochland County about Building the Said Bridge at the place Aforesaid And that they Appoint Persons to meet John Nash & Charles Anderson Gent to Agree Upon the Conditions of the Said Bridge ---

15 April 1748 O. S., Page 84
Ordered that M[r]. Nash be Appointed Surveyor of the Road that is to be Cleard from Near his house to Appomatox River a Little above Bush River ---

15 April 1748 O. S., Page 85
By an Order of Court bearing Date the 15th Day of January last Whereupon Edward Booker Richard Booker & William Booker Gent were Appointed to View the Road Petitioned for by William Kennon John Robertson & James Robertson and Make their Report to the Next Court Therefore the Said Edward Richard & William Booker Made their Report in these Words, To Wit, That the place Mentioned in the Petition for a Bridge Over Flat Creek is not So Convenient as at the Plantation of James Robertson Whereupon the Court is of Opinion that the Report is not Agreeable to the prayer of the Petition Therefore it is Ordered that no Bridge be Built at the place Aforesaid at the Expence of the County ---

15 April 1748 O. S., Page 87
Ordered that Anthony Griffin be Appointed to View the Ridge Road in the Room of William Brown who is discharged from that Office ---

20 May 1748 O. S., Page 91
Robert Ferguson Senior is Appointed Surveyor of the Road in the Room of George Evans who is discharged from that Office ---

20 May 1748 O. S., Page 91
On the Petition of Thomas Yarbrough Junior and Others for a Bridge Over Little Nottoway where Yarbrough's Road Crosses it And After Mature Deliberation being had Upon the Premisses Ordered that a Bridge be Built at the place Aforesaid And that M[r]. Irby Agree with some person or persons to Build the Same ---

20 May 1748 O. S., Page 91
On the Petition of Henry Leister for Liberty to Clear a Road Ordered that Liberty be Granted him Upon his Clearing the Same as George Moore Shall direct —

20 May 1748 O. S., Page 91
Capt. Watson Anthony Griffin & George Forster being Appointed by this Court to View the Road Near Daniel Degarnetts this Day made their Report in these Words, To Wit, That the Said Road is in good repair and very convenient as it now Stands ---

20 May 1748 O. S., Page 92
Grand Jury Presentments
... Against Charles Irby Gent Surveyor of the Road from Nottoway to West Creek ...
The Grand jury having presented Charles Irby Gent Surveyor &c And he personally Appearing It is the Opinion of the Court that he be Excused ...

20 May 1748 O. S., Page 92
Ordered that Daniel Worsham with the hands Appointed him Repair the Bridges over Smacks & Little Creeks and the Road from Bookers Fork to Andersons Road & to Bevills Bridge ---

20 May 1748 O. S., Page 92
On the Petition of Sundry the inhabitants of the Upper part of this County for a Bridge Over Buffilo River and Mature Deliberation being thereupon had It is Ordered that Charles Anderson John Nash Richard Woodson & Joseph Morton or any three of them View the Most Convenient Place where the Said Bridge Ought to be Built and Make their Report thereof to the Court ---

20 May 1748 O. S., Page 92
The hands that is to be Under the Direction of Daniel Worsham are Daniel Wilson Thomas Walters George Ragsdale Matthew Jackson Abraham Green's Quarter Capt. Worshams M^r. Towns Quarter William Callicutt and John Blanchett ---

17 June 1748 O. S., Page 93
Abraham Cocke Gent Undertakes here in Court to keep the Bridge Over Little Nottoway in Good repair Seven Years from this Day and in consideration of which he is to be paid three pounds Current Money and to Continue the road from thence to Great Nottoway ---

19 August 1748 O. S., Page 99
On the Petition of James Scott praying that he might have Liberty to Set up Swinging Gates a Cross the Road leading by his House Therefore It is Ordered that the Same be Granted him According to the Prayer of his Petition as his request Seeming to the Court to be reasonable ---

19 August 1748 O. S., Page 99
On the Petition of Samuel Cobbs Gent praying that he Might have Liberty to Set up Swinging Gates a Cross the Road leading by his House Therefore it is Ordered that the Same be Granted him According to the Prayer of his Petition as his Request Seeming to the Court to be Reasonable ---

19 August 1748 O. S., Page 100
The Persons Appointed to let the Bridge Over Appomatox a little above the Mouth of Bush river having informed the Court that they had Agreed with an Undertaker To Build the SameTherefore It is Ordered that M^r. Nash & M^r. Walker Receive the Same when done if they Shall think proper

19 August 1748 O. S., Page 101
Ordered that a Bridle way be Cleared from the Road near Charles Andersons to Bush River Church and that Joseph Morton Junior be Appointed Surveyor thereof & that Charles Andersons Richard Woodson Alexander Cunningham Theodorick Carter Joseph Shelton John Chessright's with their Male Labouring Tithables Assist the the Aforesaid Morton & be Under his Direction ---

19 August 1748 O. S., Page 101
Ordered that M^r. Nash & Charles Anderson Agree with Some Person or Persons to Build a Bridge Over Buffilloe where the Road Crosses the Same ---

19 August 1748 O. S., Page 101
Ordered that Richard Woodson with the hands Under him Clear the Road out of the Road Near M^r. Nash's to the Bridge Over Appomatox River ---

19 August 1748 O. S., Page 102
Ordered that Edward Jones be Appointed Surveyor of the Road from West Creek to the Fork of the Road Near Major Peter Jones's ---

19 August 1748 O. S., Page 102
Ordered that John Baldwin be Appointed Surveyor of the Road in the Room of Robert Vaughan who is discharged from that Office ---

19 August 1748 O. S., Page 104
James Jackson being Appointed Surveyor of a New Road that he has Cleared from the Old Road into M^r. Cocks Road Therefore It is Ordered that John Ragsdale John Hightower William Hightower Edward Jackson Joshua Morgan Tobitha & John Jones's Male Labouring Tithables be Added to the Number of hands that is Under the Said Surveyor's Direction and that they Work on the Said Road ---

19 August 1748 O. S., Page 105
Ordered that William Jackson be Appointed Surveyor of a Road from Battes Path to the County Line and that William Cryer Charles Jackson Bryant Fannell Capt Haynes's and their Male Labouring Tithables Assist him in doing the Same ---

19 August 1748 O. S., Page 105
Ordered that William Stone be Appointed Surveyor of the Road from the CountyLine into James Jackson's and that Robert Taylor Lewis Hammond William Manire M^rs. Elizabeth Poythresses

George Hill Henry Clark Robert Stadey John Bentley and all their Maile Labouring Tithables do Assist him in doing the Same ---

19 August 1748 O. S., Page 105
Ordered that Greenham Dodson be Appointed Surveyor of the Church Road in the Room of William Jackson who is discharged from that Office and that Bennedick Hammack John Simmons and John Hughes Together with all their Male Labouring Tithables Assist him in the Same ---

19 August 1748 O. S., Page 105
Ordered that Peter Wynne be Continued as Surveyor of the Road to Stokers Bridge ---

19 August 1748 O. S., Page 105
Robert Stoker has liberty to keep the Old Road Open from Peter Wynns Road to William Jacksons New Road at Bates's Path ---

20 August 1748 O. S., Page 105
M^r. Meredith having presented an Order made by Goochland County Court in which Order this Court is desired to Appoint persons to Meet the Gent Appointed by Goochland County to Agree with them Concerning the Building of a Bridge Over Appomatox river at or Near Liles's Ford Therefore it is Ordered that M^r. Scott M^r. Ford and William Archer or any two of them are Appointed to Meet the Persons Appointed by Goochland County Aforesaid to treat about the Building of the Bridge Over Appomatox Aforesaid

16 September 1748 O. S., Page 112
On the Motion of Thomas Tabb Gent It is Ordered that the Sherif pay the proportion of Genito Bridge to him out of the Money raised by the Sale of Tobacco ---

21 October 1748 O. S., Page 115
Ordered that Henry Anderson Thomas Spencer View a way for a Road Between Thomas Spencer & Edmund Booker into the Main Road from the Court house to Warwick Near the South fork of knibbs Creek and make the Report thereof to the Court ---

16 December 1748 O. S., Page 116
Henry Anderson and Thomas Spencer being Appointed in October Court last to View a Way to be Cleared from Between Thomas Spencer's & Edmund Booker Junior into the Road from the Court house to Warwick Near the South fork of Knibbs Creek and having Reported the Same to be a better way and almost two Miles Nearer to Goods Bridge than the Road that now is in Consideration thereof it is Ordered that the Road be Cleared according to the Viewers Report And that Edmund Booker Junior be Appointed Surveyor from Bush river Road above Cheathams into the Road near the South fork of Knibbs Creek and the hands that are to Work on the Said Road Under the Said Surveyor are M^r. Spencer M^r. Towns's Quarters Doctor Scott James Hill Edward Booker Junior Benjamin Wards Wilkinsons Quarters & Edward Friend George Avery and James Cheatham ---

16 December 1748 O. S., Page 116
Thornton Smith is Appointed Surveyor of the Road in the Room of Thomas Covington who is discharged from that Office which Said Road is from Cheathams to the foot of the Hill Over Flatt Creek And that William Foster James Mitchell Smiths William Wilkinson & Thomas Covington with their Male Labouring Tithables Assist the Said Surveyor ---

20 January 1748 O. S., Page 120
Ordered that Charles Clay be Appointed Surveyor of the Road in the Room of John Talley who is discharged from that Office ---

20 January 1748 O. S., Page 120
Ordered that Robert Rowland be summoned to the Next Court for his not keeping the Bridge over Deep Creek in Repair ---

20 January 1748 O. S., Page 120
On the Petition of John Robertson Ordered that he have Liberty to Clear a Road from his ford into the Road to Goods Bridge Below Peter Websters ---

20 January 1748 O. S., Page 120
Ordered that M^r. Watson Edward Robertson and Matthew Cabiness do View Yarbrough's Road and Make Report of their Opinion Upon the Premisses of the Court ---

20 January 1748 O. S., Page 120
Ordered that Alexander Roberts go and View Gennitoe Bridge and to Repair the Same if Sees that it is Wanting And bring in his Charge to the Court ---

20 January 1748 O. S., Page 124
Ordered that Major Bookers & M^r. Tarry's hands Clear the Road from Booker's Mill to the Fork of the Road at Goods Bridge And that Richard Booker Gent be Appointed Surveyor ---

21 January 1748 O. S., Page 124
Ordered that Richard Booker Junr. be Appointed Surveyor of the Road in the Room of Thomas Dunnavan who is Discharged from that Office ---

21 January 1748 O. S., Page 124
Ordered that Mr. William Bookers Hands Work on the Road Where M^r. Samuel Cobbs is Surveyor ---

28 January 1748 O. S., Page 133
County Levy
... John Tromer for painting & Lettering 47 boards at 1 lb of Tob^o. P Letter ... 994

* * *

To Joseph Scott Deced his Account for work done at Jennitoe Bridge ... £ 3

21 April 1749 O. S., Page 135
Ordered that Christopher Walthall be Appointed Surveyor of the Road in the Room of Lodwick Tanner who is discharged from that Office ---

21 April 1749 O. S., Page 136
On the Petition of Peter Wynne Ordered that a Licence be granted him to keep an Ordinary at the fork of Stokers Road ---

21 April 1749 O. S., Page 137
Ordered that the Old Road by James Cheathams be Stop'd ---

21 April 1749 O. S., Page 137
The Bridge Over Gennitoe being out of Repair Therefore it is Ordered that the same be Rebuilt and that Richard Booker William Archer Hezekiah Ford & William Booker Gent or any two of them Make Application to Goochland Court to Assist in performing the same And it is further Order'd that Richard Booker & William Archer Gent in Conjunction with the Gent Appointed by Goochland County Court Agree with an Undertaker or Undertakers for the Rebuilding the same and Compleating it in a Workman like Manner ---

21 April 1749 O. S., Page 137
Ordered that the Bridges Over Deep Creek & Sellar Creek be Rebuilt and that Richard Jones & Peter Jones Gent Agree with an Undertaker or Undertakers to Rebuild the said Bridges and it is in the Power of the Said Gent to Agree with Workmen if they think Proper for the keeping the same Bridges in Repair ---

21 April 1749 O. S., Page 137
The Bridge at Little Nottoway being out of Repair It is Order'd that Charles Irby and Abraham Cocke Gent Agree with some Workman to Repair it ---

21 April 1749 O. S., Page 137
Ordered that Edmund Booker Junior Agree with Some Person to Build or repair the Bridge near May's Over flatt Creek ---

21 April 1749 O. S., Page 137
On the Petition of Abraham Jones Junior Ordered that he have Leave for a Bridle path thro' Major Munfords Land to his Mill ---

21 April 1749 O. S., Page 137
Ordered that John Burton Gent Agree with some person to Build or Repair Burtons Bridge ---

21 April 1749 O. S., Page 138
Ordered that the Surveyor from Craddocks Bridge Upward be Summoned to Appear at the Next Court ---

21 April 1749 O. S., Page 138
Ordered that Stith Hardaway be Appointed Surveyor of the Road from Nottoway to West Creek in the Room of Thomas Burton who is Discharged from that Office ---

19 May 1749 O. S., Page 139
On the Petition of Samuel Major Ordered that he have Liberty to turn the Road round his Plantation About one hundred Yards ---

19 May 1749 O. S., Page 140
Ordered that John Davidson be Appointed Surveyor of the Road in the room of John Harden who is discharged from that Office ---

19 May 1749 O. S., Page 140
Ordered that George Forster be Appointed Surveyor of the Road in the room of James Oliver who is Discharged from that Office ---

19 May 1749 O. S., Page 140
Grand Jury Presentments
...Against the Surveyor of Liles's Ford and Flatt Creek Bridge Against the Surveyor from Flatt Creek to Southalls and from Southalls to the Court house and from Gillentens to Flatt Creek Church Against the Surveyor from Griffins Road to Bush River and the Surveyor from the Court house to the Cross Roads below M^r. Andersons and from the Said Cross Roads to the Church Against the Surveyor of the Road from Capt. Watsons to Malereys Creek

19 May 1749 O. S., Page 142
Ordered that Richard Booker Gent let the Bridge Over Flatt Creek called Burtons Bridge ---

17 June 1749 O. S., Page 156
Ordered that William Moore be Surveyor of the Road from West Creek to Irbys Road and all the Male Tithables belonging to Edwd Jones Stith Hardaway Maj^r Jones and Thomas Bottom assist to clear the Same

18 August 1749 O. S., Page 170
Ordered that the road commonly known by the name of the parsons road from Watsons road into Irby's road be a public road, and that Nathaniel Robertson be Overseer thereof, and that M^r. Watsons hands at his own house and the hands at the Quarter Whereof Mathew Cabiness is Overseer Assist in clearing and keeping the same in Repair.

15 September 1749 O. S., Page 175
Thomas Belcher having built a Bridge over Deep Creek near Major Peter Jones for which he was to be paid nine pounds nineteen Shillings & Six pence Cur^t. Money and he keeping the Same in good repair for the Space and time of Seven Years from the tenth Day of August as by the

Conditions of his Bond given for the Same more largely doth appear, it is thereupon ordered that William Watson Late Sheriff pay him the Sum aforesaid.

15 September 1749 O. S., Page 175
Ordered that William Watson late Sheriff pay unto Robert Fleming and Thornton Smith's Estate twelve Shillings and Six pence Each for repairing the uper Bridge over fflatt Creek

15 September 1749 O. S., Page 179
Ordered that Abraham Green and Samuel Cobbs Gent. do Wait on the Court of Chesterfield and desire of their Worships that the Road which leads from Goode's Bridge to Warwick may be put in good repair.

15 September 1749 O. S., Page 179
On a Motion made to the Court by Sundry Persons that the Bridge over Bush River is much decay'd and rotten upon Considering of the Same the Court is of Opinion that a Bridge higher up will be more Convenient for the passing and repassing of People the Court therefore do Order that John Nash and George Walker Gent. let the Same to Workmen who Shall build it on the lowest terms agreed on and that the Same be built at the most Convenient Place above the old one. Also that they do agree with the Workmen for the building a bridge over Sandy River at or near Hawkins's Plantation and that they receive the Said Bridges when they are Erected and finished.

15 September 1749 O. S., Page 179
Ordered that Richard Booker & William Archer Gent Receive the Bridge that is building over jeneto when the Same is finish'd

15 September 1749 O. S., Page 179
Ordered that William Baldwin be Surveyor of the Road that leads from Nottoway to West Creek in the Room of William Moore.

20 October 1749 O. S., Page 180
William Manear is appointed Surveyor of the Road in the Stead of William Stone Decd

20 October 1749 O. S., Page 180
Charles Cheatham's, John Blanchet's & Samuel Edwards's Hands are hereby Ordered to Work on William Belcher's Road.

20 October 1749 O. S., Page 181
Richard Boram is appointed Overseer of the Road in the Room of Thomas Pettis

20 October 1749 O. S., Page 181
John Clay is appointed Surveyor of the Road in the Room of Thomas Booth ju^r

20 October 1749 O. S., Page 181
Ordered that ffrancis Anderson's Road be continued to Stocks Creek.

20 October 1749 O. S., Page 181
Ordered that Capt. Watson's fflatt Creek quarters, Mallarys Creek quarter Hugh Loudon John Allbright Avery Thomas Alsup Samuel Yarborough Benj^a Bullington Work on the Road from Capt. Watson's Cart Path to Mallery's Creek and George Foster is appointed & ordered to be overseer of the Same.

20 November 1749 O. S., Page 184
On the Petition of Sundry Inhabitants in the fork of Buffelow praying that a Road may be Cleared from William Watson's into the Road at Brathwait's the Court Considering the Same it is ordered that the Said Road be Cleared and that Edward Braithwait be Surveyor of the Same

20 November 1749 O. S., Page 185
Grand Jury Presentments
...The Surveyor of the Road on both Sides of Ferguson's Bridg, The Sellers Creek Bridge & Conway the Surveyor, the Surveyor of the Road from Stocks Creek to Samuel Majors, The Surveyor of the Road from Burton's Bridge to the Church, ... the Surveyor of the Road from Browne's Ordinary to Malary's Creek.

20 November 1749 O. S., Page 186
Mr. Green being Some time past appointed to View the Lower Bridge over Deep Creek and the Beaver Pond Creek and having reported that the Same are insufficient and Very dangerous for People to Pass over them it is therefore ordered by the Court that M^r. Green and M^r. Tarry do let the Said Bridges to Workman who Shall undertake to build the Same at the lowest price

20 November 1749 O. S., Page 186
Thomas Carter is appointed Surveyor of the High Way that leads from the upper fork of Sandy River to George Moore's in the Room of John Hardin

18 January 1749 O. S., Pages 189-190
County Levy
To Daniel Coleman for building a Bridge over Wintycomeck ... 500 [lbs Tobacco]

* * *

To Peter Clarke for building a bridge over the Cellar Creek - Cash   £ s d. ... 5..0..0

* * *

To Thomas Walberton for taking care of Goode's Bridge from the year 1746 ... 6..14..0

To John Clarke for the Bridge over Deep Creek near Burton's for Seven Years ... 12..0..0

To the Same person for a bridge over the Beaver pond branch ... 3..14..6

To Abraham Hart for building a Bridge over fflat Creek ... 6..10..0

To John Thomas for building a bridge over little Nottoway ... 3..17..6

To William Towns for labouring to Secure Jeneto bridge ... 0..13..0

* * *

To Pay Alexander Roberts leveyed on the County for building of a bridge over Appomattox River & the other Money Debts before mentioned ... 10903 [lbs Tobacco]

19 January 1749 O. S., Page 191
William Ray is appointed Surveyor of the Road which leads from Saylors Creek to Sandy Creek.

19 January 1749 O. S., Page 192
Ordered that John Towns with the hands under him do clear the Road which leads from Jeneto Road to the new Bridge.

19 January 1749 O. S., Page 192
Ordered that Charles Irby Gent. agree with Some Person to build a Bridge over West Creek at Such Place as he Shall See that will be most Convenient.

19 January 1749 O. S., Page 192
Ordered that a Road be Clear'd from the Head of Yarborrough's Road up the Ridge between Peters Creek and Whetstone to Tukaer's Cart Path thence along the Path up the Ridge between Little Nottoway and great Nottoway into the Road near Degernetts, and that John Payne be Surveyor of the Same and that James Callicoat James Prisnall, Ralph Shelton William Lees qr^t. William Quin Thom^s. House, Peter Dupuy ju^r, Thomas Chandler and Richard Cruchfield assist him in Clearing the Said Road

19 January 1749 O. S., Page 193
Ordered that Kennon's Hands, be added to John Towns's Gang and M^{rs}. Scotts to M^r. Walker's and that M^r. Walkers Hands do Assist Towne's Gang in Clearing the New Road to the Bridge ---

16 February 1749 O. S., Page 197
Ordered that the Road from Appomattox River to the County Line be clear'd and that the following hands Assist James Nix who is appointed Surveyor of the Same (to wit) James Doss, Elisha Lyan, Edward Nix's Tythes, John May, Henry Dicke's, Noel Gibson and Josiah Payne.

16 February 1749 O. S., Page 198
On the Petition of John Pride Setting forth that the Road which goes through his plantation is very hurtful to him and therefore prays leave that he may have liberty to turn the Same to his advantage the Court Considering of the same grants him leave according to his petition

16 February 1749 O. S., Page 199
Ordered that the Surveyor of Saylors Creek Road turn the Same opposite to Charles Johnson's the best way he can into the Road that leads from Craffords to Dawsons Race paths.

16 February 1749 O. S., Page 199
Pursuant to an Order of this Court M^r. Hezikiah Ford has agreed with Robert Ferguson to build a bridge over Flatt Creek and to keep the Same in Good Repair for Seven Years after the twenty fifth of March next for which the Said Ferguson is to have ten Pounds paid to him in September next if the Same be finished.

16 March 1749 O. S., Page 200
John Compton is appointed Surveyor of Lyles's Road in the room of William Hutchinson.

16 March 1749 O. S., Page 201
On the Motion of Thomas Williamson leave is given him to Clear a Road from Saylors Creek to M^r. Nashes Road above Womawks the most Convenient Way that he can find, Ordered that M^r. Cobbs, M^r. Towns, Maj^r Booker's W^m. Sadler's Willm Liggon's Abraham Jones's & James Coleson's Male Tythes assist in Clearing the Same Thomas Williamson is appointed Overseer of the Same

16 March 1749 O. S., Page 202
On the Motion of William Ray; the illconveniency of the Road near his plantation leave is therefore granted him to Clear it round his fence

16 March 1749 O. S., Page 203
Leave is given to Clear a Bridle Way from or near the Head of Snails Creek into Bush River Road that leads to Sandy River Chappel

16 March 1749 O. S., Page 205
Ordered that William Southal and his Tythes be added to William Fergusson's Gang

16 March 1749 O. S., Page 205
Ordered that George Booker, Hezekiah Ford, M^r. Anderson and John Chistam do assist William ffergusson with all the hands under them to meet at flatt Creek Bridge and there make the Causeys to the Same.

16 March 1749 O. S., Page 205
Thomas Foster is appointed Surveyor of the Road that leads from Stocks Creek to Sandy Creek in the Room of John Mauldin and that George Forster Joel Meadows, Joseph Pollard, John Forster and John Morris ju^r be added to his Gang ---

21 April 1750 O. S., Page 224
On the Motion of Abraham Hurt representing to this Court that he has built the bridge over Flatt Creek and that the Same is perform'd in a much better Manner than was agreed for he therefore Prays for a further allowance for the Same and the Court being made Sensible that his representations were true and just it is therefore ordered that a further Sum of three Pounds be allow'd him for his faithful Performance and that the Sheriff do pay him the Same.

21 April 1750 O. S., Page 224
Ordered that Mr Green and M^r. Tarry Receive the Bridge that is built Over Beaver Pond Branch if they think it is built agording to agreement, and that they agree with Some Person to keep the Same in Repair for the Space of Seven Years

21 April 1750 O. S., Page 226
Hezekiah Ford Gent came into Court and made the following Report Persuant to an Order directed to him for that Purpose, that he had let the Bridge over Flatt Creek to an undertaker, that Robert Fergusson had agreed with him to keep the Same in Repair for the Space and time of Seven Years for which he is to pay him ten pounds and deliver up the Said Bridge the Eighth Day of May it is thereupon Ordered that the Sheriff Pay M^r. Ford ten pounds and that he receive the Said Bridge if he thinks it proper

18 May 1750 O. S., Page 234
Ordered that William Browne be added to John Martin's Gang and John Dawson and that any two of them View a Way for a Road that was Ordered last Court and return their Opinions thereon to the next Court.

18 May 1750 O. S., Page 234
Francis Green is appointed Surveyor of a Road that leads from Anderson's Bridge down in the Room of Robert Taylor and that the Same hands under Taylor assist in clearing the Same

18 May 1750 O. S., Page 236
Ordered that Charles Irby and Abraham Cocke Gent meet the Gentlemen appointed by Lunenburg Court at Dyer's lick in order to agree with Workmen to build a bridge over Nottoway River at or near the Said lick.

18 May 1750 O. S., Page 236
Grand Jury Presentments
... the Surveyors of the Road from Flatt Creek to Lilles's Ford.

* * *

Col^o. William Randolph's Overseers for fencing the High way up

15 June 1750 O. S., Page 243
George Moore is appointed Overseer of the Road from his House to the Hatters And it is Ordered that the former Hands assist him in keeping the Same in Repair

15 June 1750 O. S., Page 245
Ordered that a Road be clear'd from James Atwoods Road the Nearest and best Way into Roanoak Road and that Atwood's Tythes, Flournoy's Tythes Abraham Eastis, John Popham, Robert & Samuel Holderness and their Tythes Assist Abraham Eastis, Surveyor of the Said Road to Clear the Same

15 June 1750 O. S., Page 245
Ordered that Gardiner Mayes be Surveyor of the Road from flatt Creek Bridge at Mayes's to the fork of Bush River Road near Walters's Road in the Room of Thornton Smith and the Same hands to assist him in Clearing the Same.

15 June 1750 O. S., Page 245
William Ewing is appointed Surveyor of a Road from Sawneys Creek into Randolph's Road and that Joel Watkins & c^a. to Work on the Same

21 July 1750 O. S., Page 263
Ordered that the Sheriff pay to Thomas Belcher five pounds ten Shillings for the building of a bridge over West Creek

19 October 1750 O. S., Page 273
Daniel Tucker is appointed Surveyor of the Road that leads from Namozeen Creek to Coles Branch in the Room of Charles Clay

19 October 1750 O. S., Page 276
Ordered that Thomas Tabb & William Archer Gent receive the Bridge newly buit over jenito if it is Perform'd according to agreement, also that they agree with Some Person to take care of the Said Bridge and that they Solicit the other County to bear their Proportionable Expences in keeping the Said

19 October 1750 O. S., Page 276
Ordered that Capt. Jones agree with Some Person to Rail in Winticomake Bridge which is at Present Somewhat dangerous to Pass without Such Railing

20 October 1750 O. S., Page 279
Ordered that John Nash Gent be Surveyor of the Road that's to be clear'd from the Race Paths at Abraham Wawmocks to Sandy River Bridge from thence to Bush River Bridge and into the Buffelow Road, And that George Walker's Tythes Henry Liggon's Abraham Wawmock's, Thomas Haskins's Jonathan Cheatham's, William Lewis's and Douglass Puckett's Tythes assist him in Clearing the Same.

20 October 1750 O. S., Page 279
Ordered that Joseph Morton and Charles Anderson agree with workmen to build bridges over Bush River and Bryary River at the most convenient Places for the Same and make return thereof to this Court.

20 October 1750 O. S., Page 279
Ordered that Abraham Green agree a Workman to build a bridge over Wintocomack were the lower Road Crosses and make return thereof to this Court

16 November 1750 O. S., Page 289
On the motion of John Twitty, Ordered that Robert Stoker & John Bridgforth View the Roads and make report which is the most Convenient Way from the Said Stokers to Clayborn's to the next Court

16 November 1750 O. S., Page 289
Ordered that William Crawley view the Road and report its true conveniency as Wood Jones Gent. Was appointed to View

16 November 1750 O. S., Page 289
Ordered that James Wimbush be Surveyor of the Road that leads from Charles Anderson's to the Extent of the County towards little Roanoak Bridge and that Caleb Baker, Hugh Nixon, John Bibbs, John Ritchey George Davis James Thaxton, James Atcherson, Patrick Galaspie, John Cox and the Said Wimbush's Tyths assist in Clearing the Same.

16 November 1750 O. S., Page 289
Grand Jury Presentments
... the Several Surveyors of the Roads from Capt. Watson's to the fork above George Moore's

16 November 1750 O. S., Page 292
Ordered that William Bass ju^r Christopher Bass and John Man's Male Tythes be added to John Towns's Gang

16 November 1750 O. S., Page 292
ordered that the Hands of John Pain work under Daniel Dejarnett till he has put his Road in Good Order

16 November 1750 O. S., Page 292
Orderd that Robert Atkins Surv^r. of the Road from George Moore's to Sandy River and all the Tythes on the North Side of the Road Except Coles Assist him in Clearing his Road

18 January 1750, Page 298
Ordered that John Clay be appointed Surveyor of the Road in the Room of Thomas Booth

18 January 1750 O. S., Page 298
Ordered that William Barnes be Overseer of the Road that leads from Sandy Creek to the Fork of the Road that leads to Jeneto in the Room of Thomas Whitworth

18 January 1750 O. S., Page 299
The Bridge built by Thomas Anderson over Appomattox River at a place known by the name of Bevills being by the innundation of Water carried away the Said Anderson Appearing this Day refused and Denyed to Rebuild the Same According to the Conditions of a Certain Writing Obligatory which Specified that he Should Maintain the Said Bridge for Seven years Completely, The Court thereupon Ordered that William Booker Abraham Green and Wood Jones Gent. or any one of them do Acquaint the Court of Chesterfield of Such Refusal and to desire them to appoint one or more of their Members in Conjunction to agree to build the bridge aforesaid at the Place aforesaid

18 January 1750 O. S., Page 300
Ordered that James Oliver be overseer of the Road in the Room of Robert Moody.

18 January 1750 O. S., Page 300
Ordered Alexander Bruce be Overseer of the Road in the Room of Charles Irby Gen^t. that Leads from Nottoway Church to West Creek

18 January 1750 O. S., Page 300
Thomas Tabb, M^r. Booker & William Archer Gent. are appointed to treat with the Court of Cumberland County about and Concerning the building of a Bridge at or near Geneto and to desire the Said Court of Cumberland to appoint one or more of their Members in Conjunction to agree with Some Workman for the build of the Same

18 January 1750 O. S., Page 300
Ordered that Samuel Cobbs Gent. Surveyor of the Road which leads from the Church to Ferguss's Bridge Warn the Several Tythes that Works on the Same together with the tythes at Anderson's Quarter Robert Fergusson Senr.'s Tythes William Southal's Tythes, James Fergusson's Tythes William Fergusson's Tythes George Evans's Tythes & Benjamin Hubbard Meet on the Said Road in Order to make a Causey on the uper Side of the Said Bridge

15 March 1750 O. S., Page 301
Ordered Catlet Man be Overseer of the Road in the Room of Henry Ward Warn the Several Tyths work on the Same together with the tyths of Mr Estis Charles Hutcherson Tyths Walter Mitchells Elisha Estis Junr. tyths Alexander Marshall Tyths at his Quarter and Stith Hardways tyths at his Quarter and John Worsham tyths at his Quarter Edwd. Taylor Henry Wards tyths and Catlets Mans Tyths Assist in Clearing the same.

15 March 1750 O. S., Page 303
Ordered that Daniel Dejarnette turn the Road round the Head of Mallory's Creek the best & most Convenientest Way he can find.

15 March 1750 O. S., Page 304
Richard Ellis is appointed Surveyor of a Road that is to be Cleared from the Road by Dandy's into the Road Near Phillip Pledger's and that John Ellis, Hampton Wade and William Dandy Assist in Clearing the Same

15 March 1750 O. S., Page 304
On the Motion of William Mayes Surveyor of the Road from Flatt Creek to Crawford's House It is ordered that James Foster Abraham Forrest, Benjamin Hawkins George White their Tythes and the Tythes of the Said Mayes Assist in Clearing the Same

15 March 1750 O. S., Page 305
Ordered that Edward Brathwett and the People under him Clear A Road out of Randolphs Road below his lower Quarter to the upper Church in Nottoway Parish

15 March 1750 O. S., Page 305
Thomas Tabb Hezekiah Ford Gent. and Benj^a. Harris or any two of them are appointed and Desired to treat with the Court of Cumberland about Repairing the upper Bridge Over Appomattox River and George Walker and Charles Anderson are appointed by this Court in Conjunction with those that are appointed by the Court of Cumberland to agree with a Person or Persons to Repair the said Bridge.

15 March 1750 O. S., Page 306
Ordered that Charles Anderson and Joseph Morton agree with Some Person for the repairing the Bridge over Buffeloe River

15 March 1750 O. S., Page 306
Ordered that Charles Anderson and Joseph Morton agree a Workman or Workmen to build a bridge over Mountain Creek at or near where the Road Crosses the Same

15 March 1750 O. S., Page 307
Ordered that Richard Booker Abraham Green and Wood Jones Gent or any one of them meet in Conjonction with the Gentlemen of Chesterfield County and agree with Workmen to build a bridge at or near Burton's over Appomattox River

15 March 1750 O. S., Page 307
Ordered that John Smith the Widow Smiths Tythes, M^{rs}. Mitchel's Tythes and William Williamson & Thomas Hawkins Work on the Road that Gardiner Mayes is Surveyor of (to witt) from Flatt Creek Down to Bush River Fork.

15 March 1750 O. S., Page 307
Thomas Tabb & William Archer Gent Persuant to an Order of this Court appointing them in Conjunction with Some of the Members of Cumberland Court to agree with Persons for the building a bridge over Appomattox River at or Near Geneto made the following report that they had agreed with Joseph Epperson and John Elliot to build the Said Bridge and to keep the Same Passable for Carts and other Wheel Carriages for the Space and term of Eleven Years four Months and one Day from the time the Same Should be finished untill the aforesaid term and time Should be fully expired, and for which they are to receive one hundred & Eighty Pounds Current Money to be paid to them in the Month of August which Shall be in the Year of our Lord one thousand Seven Hundred and fifty two and also to be Paid after the Expiration of the Said Eleven Years four Months and one Day five Pounds for Every Year After for keeping the Same in repair

15 March 1750 O. S., Page 308
ffrancis Anderson and William Liggon ju^r are appointed Surveyors of the Appomattox River Road that leads from Jeneto to Bush River and to View the Same and report their Opinion of it to May Court

19 April 1751 O. S., Page 318
George Elliott is appointed Surveyor of the Road in the Room of Dacey Southall

19 April 1751 O. S., Page 318 (319)
On the Petition of of Benjamin Hendrake Leave is granted him to Clear a bridle Way through the Lands of John Farley Isaac Morris and Stewart Farley into the Main Road the most Convenient Way to the Church So as not to injure their Plantations

19 April 1751 O. S., Page 318 (319)
Ordered that Mark Jackson and Thomas Jackson work on the Road that William Mayes is Surveyor off

17 May 1751 O. S., Page 321
On the Petition of Sundry the inhabitants of upper Part of this County Setting forth the great Necessity of a bridge being built over Appomattox River at or near Sandy Ford and the great inconveiniances and hardships the lye under for the want of Such a bridge the Court on Considering the Allegations of the Said Petition Depute John Nash and George Walker Gent two of their Members to view the Place aforesaid and to make report thereof to the next Court.

17 May 1751 O. S., Page 322
Charles Irby and Abraham Cock Gent. are appointed in Conjunction with Richard Whitton and Mr. Lawson two of the Members of Lunenburg Court to let the bridge that is to be built over great Nottoway at the most Convenient Place that will be Answerable to Cocks Road to Such Person or Persons as will undertake to build the Same

17 May 1751 O. S., Page 322
Simcock Cannon Ellis Palmer Isaac Chandler and John Morton are Ordered to Assist in Clearing a Road from Atwoods Plantation on Bryery River to Roanoak Road under Ambrose Eastis Surveyor of the Same

17 May 1751 O. S., Page 322
Ordered that George Moore and Daniel Dejarnett with the Hands under them turn the Road along the Ridge from Moore's below Dejarnett's Smith Shop and Moore is to come no lower than the Head of Snails Creek

17 May 1751 O. S., Page 322
Joseph Morton is appointed to Clear the Road from Abram Bakers Mill Path to the upper Botton that goes through Plantation and M^r. Wimbushes from thence to the County Line that Capt Andersons hands and John Martins hands do Assist the said Morton with his hands in Clearing the Road

17 May 1751 O. S., Page 323
Daniel Jones has Leave to turn the Road round his fence if he puts the Same in Good Repair as before.

17 May 1751 O. S., Page 323
Ordered that Martin Wilkinson be Survey^r. of Nottoway Road to west Creek in the Room of William Baldwin

17 May 1751 O. S., Page 323
Ordered that William Cross be Surveyor of the Road from little Nottoway to Where the New Bridge is to be built and that Abraham Cocke's Hands William Cross's Thomas William Jones, Moses Hurts & George White Assist him in Clearing the Same

17 May 1751 O. S., Page 323
Ordered that William Mayes be Surveyor of the Road from little Nottoway Bridge in Jordan's Road to William Maynard's and Alexander Bolling William and Henry Batts Peleg Furgusson and William Red and all their Male Tythes Assist him in Clearing the Same.

17 May 1751 O. S., Page 323
Abraham Green is Appointed and desired to Receive Wintocomeck bridge

17 May 1751 O. S., Page 324
Ordered that the Sheff. Pay John Nash Gent Sixteen Pounds ten Shillings Current Money for building the bridge over Bush River

17 May 1751 O. S., Page 324
Grand Jury Presentments
... The Surveyors of the Road from Watson's to the Head of Bryery River

* * *

The Surveyor of the Road from the Fork above Baldwinn's Ordinary to James Cheathams
The Surveyor of the Road from Goode's Bridge to Tanner's Road thence to Anderson's Road and to Thomas Jones's
The Surveyor of the Road from George Moore's to Sandy River

* * *

John Clarke for not keeping the lower Bridge over Deep Creek in repair
... the Surveyor of the Road from Sandy River to Nash's Mill on Bush River.
... the Surveyor of the Road from Watson's Road to Nottoway Church
... the Surveyor of the Road from Watson's Race Ground to the fork below Daniel Jones's

17 May 1751 O. S., Page 327
Ordered that Charles Anderson View and Receive the Bridges built over Buffelow River & Mountain Creek if he thinks the Same Sufficient and to make report thereof at the next Court

17 May 1751 O. S., Page 330
Ordered that John Hurt be appointed Surveyor of the Road in the Room of Thomas Anderson from Stokes's Creek to Samuel Majors and that all the Persons hereafter mentioned Assist him in Clearing the Same (to Wit) Pauling Anderson Gesse Seay, William Fosters, M^r. Walker's Quarter Tythes M^r. Ford and Samuel Ellin With all their Male Tythes to Work on the Same

18 May 1751 O. S., Page 332
Abraham Cock and Leonard Clayborn, Gent. are appointed to apply to the Court of Brunswick to Rebuild the Bridge at Stoker's Over Nottoway River which is fallen to decay and Abraham Cock in conjunction with the Gent. Appointed by the Court of Brunswick is to agree with Some Workman or Workmen to build the Same

25 July 1751 O. S., Page 341
John Taylor is appointed Surveyor of the Road from Nottoway Chappell to the County line Same Land As Usual

25 July 1751 O. S., Page 341
Alexander Bruce is appointed Surveyor of the Road from Nottoway Church along ye Courthouse Road to west Creek ordered that the Same hand work thereon

25 July 1751 O. S., Page 341
William Evans is appointed Surveyor of the Road from Great Nottoway at Hampton Wades to Nottoway Bridge the Same hand work thereon

25 July 1751 O. S., Page 341
John Thomas is appointed Surveyor of the Road from Jordan:s Bridge to M^r. Cock:s Road

25 July 1751 O. S., Page 341
Arthur Leath is appointed Surveyor of the Road from M^r. Cocks Road to Nottoway Road ---

25 July 1751 O. S., Page 341
William May is appointed Surveyor of the Road from Thomas:s Road to Little Nottoway Bridge.

25 July 1751 O. S., Page 341
William Cross is appointed Surveyor of the Road from Great Nottoway Bridge to Little Nottoway

25 July 1751 O. S., Page 341
Peter Wynn is appointed Surveyor of the Road from Stokers Bridge to Jackson:s Road

25 July 1751 O. S., Page 341
Robert Farguson is appointed Surveyor of the Road from Nottoway bridge to the Harricane Bridge ...

25 July 1751 O. S., Page 342
Greenham Dodson is appointed Surveyor of the Road from the Harricane Bridge to the County Line---

25 July 1751 O. S., Page 342
James Anderson Senr. is appointed Surveyor of the Road from Leigh's Bridge to the old Road below his House and the Hands under the old overseer to assist him

25 July 1751 O. S., Page 342
All the Hands that are convenient which formerly work'd on the Old Road are to be under the Said Anderson.

25 July 1751 O. S., Page 342
Richard Ellis is appointed Overseer of the Road which leads from out of the Road by Dandy's to Philip Pledger's

25 July 1751 O. S., Page 342
John Pain is appointed Surveyor of the Road from Richard Stone's to the head of Buffelow-Bed Creek

25 July 1751 O. S., Page 342
Nathaniel Robertson is appointed Surveyor of a Road that leads from the Court House Road to Griffins Road.

25 July 1751 O. S., Page 342
George Foster is appointed Surveyor of the Road that leads from the fork below the Race Paths at Watson's to Mallery's Creek

25 July 1751 O. S., Page 342
Daniel Dejarnette is appointed Surveyor of the Road that leads from Mallary's Creek to Snails Creek.

25 July 1751 O. S., Page 342
Anthony Griffin is appointed Surveyor of the Road that leads from Snails Creek to George Moore's.

25 July 1751 O. S., Page 342
James Oliver is appointed Surveyor of the Road that leads from the Race Paths to the Court House Road.

25 July 1751 O. S., Page 342
Cattlin Mann is appointed Surveyor of the Road that leads from West Creek to the Court House.

25 July 1751 O. S., Page 342
Martin Wilkinson is appointed Surveyor of the Road that leads from the Cross Road to West Creek

25 July 1751 O. S., Page 342
Edward Jones is appointed Surveyor of the Road that leads from West Creek to Deep Creek

25 July 1751 O. S., Page 343
George Booker is appointed Surveyor of the Road from Edmund Walkers to the Road by Nicholas Guilintines

25 July 1751 O. S., Page 343
John Chissum is appointed Surveyor of the Road that leads from Guilintine's new Road to the Road that comes Down by Burton's.

25 July 1751 O. S., Page 343
Hezekiah Ford Gen^t. is appointed Survey of the Road from the Causey of Fargusons Bridge to M^r. Clement:s

25 July 1751 Old Style, Page 343
David Flournoy is Appointed Surveyor of the Road from the head of Pole Bridge Down to the fork his hand Mat Flournoy Washbuns hand to work on the Same.

25 July 1751 Old Style, Page 343
Edmund Booker Jun[r]. is appointed Surveyor of the Road from the Old above Jam[s]. Cheathams to Knibbs Creek

25 July 1751 Old Style, Page 343
William Foster is appointed Surveyor of the Road from Baldwins to the fork above Cheatham:s John Smith's hand Assist in Clearing the Same.

25 July 1751 Old Style, Page 343
Francis Anderson from Stocks Creek to Gullingtons.

25 July 1751 Old Style, Page 343
John Hurt is appointed Surveyor of the Road from Stock Creek to Samuel Major:s

25 July 1751 Old Style, Page 343
John Cumpton is appointed Surveyor of the Road from Lisles fford to fflatt Creek Bridge

25 July 1751 Old Style, Page 343
William Mayes is appointed Surveyor of the Road from Crawfords house to fflatt Creek Bridge.

25 July 1751 Old Style, Page 343
Gardner Mayes is appointed Surveyor of the Road from flatt Creek Bridge to the fork below Fosters

25 July 1751 Old Style, Page 343
Edward Harper is appointed Surveyor of the Road from Watson Race Ground to Baldwins

25 July 1751 Old Style, Page 343
John Clay is appointed Surveyor of the Road from Winticomak to Spinners Run

25 July 1751 Old Style, Page 343
Dan[l]. Tucker Surveyor of the Road from Namazeen Bridge to to the fork of Greens Road

25 July 1751 Old Style, Page 344
Robert Tucker is appointed Surveyor from Green's Road To Wintocomeck Bridge

25 July 1751 Old Style, Page 344
Robert Cozins is appointed Surveyor of the Road from the fork of Green's Road to a Chesnut Oak.

25 July 1751 Old Style, Page 344
Thomas Green Sen[r]. is appointed Surv[r]. of the Road from a Chesnut Oak to Deep Creek lower Bridge

25 July 1751 Old Style, Page 344
Thomas Bevill is appointed Surveyor of the Road from Deep Creek lower Bridge to the fork of Bookers Road above Borams & thence to bevills Bridge

25 July 1751 Old Style, Page 344
Thomas Green ju^r is appointed Surveyor of the Road from Bookers fork to Andrews branch

25 July 1751 Old Style, Page 344
Daniel Worsham is appointed Surveyor of the Road that leads from the fork of Bookers Road to the fork of the Road thence to Goode's bridge

25 July 1751 Old Style, Page 344
Christopher Walthall is appointed Surveyor of the Road that leads from Andrews branch to Nibbs Creek and thence from the fork below Anderson's to Bookers Road.

25 July 1751 Old Style, Page 344
Thomas Pettis is Appointed Surveyor of the Road that leads from flatt Creek Bridge on Lisles's Road to M^r. Cobbs's Ordinary.

25 July 1751 O. S., Page 344
George Elliott is appointed Surveyor of the Road that leads from the fork below the Ordinary to Bookers Mill.

25 July 1751 O. S., Page 344
Richard Booker Gent is appointed Surveyor of the Road that leads from the Mill to the fork at School House.

25 July 1751 O. S., Page 344
Richard Booker ju^r is appointed Surveyor of the Road that leads from Nibs Creek Bridge to Murray's Path.

25 July 1751 O. S., Page 344
Benjamin Hawkins is appointed Surveyor of the Road that leads from the Court House to Cobbs Ordinary.

25 July 1751 O. S., Page 344
Peter Webster is appointed Surveyor of the Road from Murray's Path that leads to Goode's Bridge

25 July 1751 O. S., Page 345
Edmund Walker is appointed Surveyor of the Road that leads from his House to Stony Bridge.

25 July 1751 O. S., Page 345
John Burton Gent is appointed Overseer of the Road that leads from his House to the Court House.

25 July 1751 O. S., Page 345
John Towns is appointed Surveyor of the Road that leads from Stony Bridge to jeneto Bridge.

25 July 1751 O. S., Page 345
Samuel Cobbs Gent is appointed Surveyor of the Road that leads from Flatt Creek at Burton's Bridge to Smiths Creek bridge at Winterham & from James Ferguson to the Church

25 July 1751 O. S., Page 345
Samuel Goode is appointed Surveyor of the Road from that leads from Sailors to Sandy Creek

25 July 1751 O. S., Page 345
William Barnes is appointed Surveyor of the Road that leads from Sandy Creek to the Folly.

25 July 1751 O. S., Page 345
Thomas Whitworth is appointed Surveyor of the Road that leads from th Folly to the fork above Burton's with the Same hands as were und Farley.

25 July 1751 O. S., Page 345
Peter Davis is appointed Surveyor of the Road that leads from George Moore's to the fork of Randolphs Road.

25 July 1751 O. S., Page 345
Robert Atkins is appointed Surveyor of the Road that leads from George Moore's to Sandy River

25 July 1751 O. S., Page 345
John Roberts is appointed Surveyor of the Road that leads from the Court house to the fork below Anderson & to have the Same hands as were under Charles Burks

25 July 1751 O. S., Page 345
Richard Jones is appointed Surveyor of the Road that leads from the Court House Road to West Creek

26 July 1751 O. S., Page 347
David Flournoy is appointed Surveyor of the Road from Bryery River to George Moore's in the Room of Peter Davis and the Same hands to assist him that assisted Davis

22 August 1751 O. S., Page 349 (Note: Date appears to be 28 in book but subsequent days are the 23rd and 24th)
On the Petition of Anthony Griffin it is ordered that his Tythes work on the Road where he is overseer.

22 August 1751 O. S., Page 350
John Burton Gent. is appointed Surveyor of the Road that leads from Jeneto Road to the Court House and that all the hands under Farlow & the Said Burton Assist him in Clearing the Same

22 August 1751 O. S., Page 350
Ordered that James Nicks be appointed overseer of the Road from the Ridge at the Head of Appomattox River joining of Callaway's Road the best Way to where it Crosses the Said River

22 August 1751 O. S., Page 351
Ordered that James Walker be appointed Surveyor of the Road that leads from Vaughan's Creek to Randolph's Road.

22 August 1751 O. S., Page 351
Thomas Haskins is Appointed Surveyor of the Road that leads from Sailors Creek Bridge to the fork of the Road a little above Sandy River Bridge and that his Tythes Abraham Wawmock's John Gentry, Charles Causton, John Spradling, Benjamin Ruffin's Tythes, Thomas Wawmock Abraham Wawmock ju^r and John Nash's Tythes on Sailor's Creek, this Gang and William Wawmock's Gang are to meet and repair Sailors Creek Bridge, And then from the Fork above Sandy River Bridge to Bush River Bridge and from thence to Buffelow Road M^r. Nash's Hands is to Clear the Same and William King is appointed Surveyor of the Said Road

22 August 1751 O. S., Page 351
Ordered that Leonard Cheatham, M^{rs}. Stokes, Thomas Tabb's Gent. and Richard Echols's Male Tythes Work on the Road that George Elliott his Surv^r. of

22 August 1751 O. S., Page 352
William Wawmock is appointed Surveyor of the Road that leads from Crawfords House to great Sailors Creek Bridge and that the Same hands as before Assist him in Clearing the Same also to Assist Thomas Haskins to repair the Said Bridge

22 August 1751 O. S., Page 352
Henry Liggon is appointed Overseer of the Road Which leads from the Fork above Sandy River Bridge to the Fork above George Walkers and that the Same hands as before assist him in Clearing the Same

22 August 1751 O. S., Page 352
William Towns is appointed Surveyor of the Road in the Room of John Towns

22 August 1751 O. S., Page 352
Douglass Puckett is appointed Surveyor of the Road that leads from Bush River Bridge to the Church and that the Same Hands usual Assist him in Clearing the Same

22 August 1751 O. S., Page 352
Alexander Frazier is appointed Surveyor of the Road in the Room of John Davidson from the Church to the old Road

22 August 1751 O. S., Page 352
Benjamin Woodson is appointed overseer of the Road that leads from the fork to Rutledge's Ford in Appomattox River above the Mouth of Bush River

22 August 1751 O. S., Page 353
Hantz Hendrake is appointed overseer of the Road from Crawford's House to the Folly and that the Same Hands as usual Assist him in Clearing the Same

22 August 1751 O. S., Page 353
Ordered that William Thurman, Richard Thurman, John May, Richard Bennett, James Bennett, James Nicks, James Doss, John Shinnaison, William Roberts, Valentin Gibson, Elisha Lyon, Edward Nicks, James Lax, Benjamin Nicks, William Nicks and Valentine Nicks Work on the Road that James Nicks is Surveyor of.

22 August 1751 O. S., Page 353
Ordered that the Sheriff pay to Daniel Coleman for his building of a Bridge over Witocomeck five Pounds ten Shillings current Money he having deliver'd the Said Bridge to Abraham Green Gent who was Deputed to receive the Same

22 August 1751 O. S., Page 353
Ordered that the Sheriff pay to Walter Mitchell three Pounds for building a bridge over West Creek on the Court House Road

24 August 1751 O. S., Page 363
A Petition of Divers of the Inhabitants of the upper Part of this County was this Day read Setting forth the great Necessity and Conveniancy of a Bridge to be built over Appomattox River at or near a place Commonly known by the name of Sandy Ford the Court Considering the Reasonableness of the Petitioners Depute Richard Booker and Thomas Tabb Gent two of their Members to treat with the Court of Cumberland County of and concerning the building the Said Bridge and to desire of them to appoint Some of their Members or Some other person or persons in conjunction with them to agree with Workmen about the building thereof

26 September 1751 O. S., Page 369
William Hall is appointed Surveyor of the Road from the fork of Anderson's Road down to Websters commonly called Tanners Road and that the Same hands which work'd heretofore on the Said Road Assist him in Clearing the Same.

26 September 1751 O. S., Page 371
Richard Jones ju^r is appointed Surveyor of the Road which leads from the Court House Road to West Creek Road below Hudson's Race Paths and the Same Hands as Work'd on the Same before Assist him in Clearing the Same.

26 September 1751 O. S., Page 371
Ordered that James Walker be Surveyor of the old Roling Road that leads out of Randolphs Road near his Mill and thence across Vaughan's Creek at the Old Ford and then to Continue the old Way to the mouth of Sawneys Creek and that Charles Simmons, Manase Mackfield, William Hill, Robert Jennings, Arthur Neil, Giles Evans, Samuel Matthews, Peter Coffee, John Anderson, Andrew Mackadoo, Robert Blake, James Blake, Andrew Dow, Robert Martin, William Smith, Lawrence Moreau, Robert Forbush, Manaseh Mackbride, William Ewing, John Morrow, Thomas Ewing, James Daniel, Thomas Fulton, John Caldwell, James Mackew, James Armstrong, Robert Gresham, William Miller, William Kelley, Thomas Stone, Henry Childs, John Phips, John Hannah, James Ewing, William Macklew, George Ewing, George Coock, James Walker, Elkanah Jennings, John Childs, and their Several Male Tythes to assist and Work on the Said Road.

27 September 1751 O. S., Page 373
Daniel Coleman having Six pounds ten shillings due to him from this County for the building of a bridge over Wintercomeck as also Several other persons Sundry other Sums for public Edifices Errected by them and the County's fund being not at present in a Sufficiency to discharge its Several Debts It is therefore ordered that Lawful Interest be allowed Each person untill their Several Demands be duly Satisfied

24 October 1751 O. S., Page 382
On the Motion of George Steegall leave is given him to Clear a bridle way from his House to the Church Road & c^a.

24 October 1751 O. S., Page 383
Ordered that Thomas Booker ju^r be Surveyor of the Road in the Room of John Clay

24 October 1751 O. S., Page 383
Ordered that Abraham Green and Wood Jones Gent agree with Workman or Workmen to build a Bridge over Sweathouse Creek at or near where the Road crosses the Same

24 October 1751 O. S., Page 386
Ordered that Thomas Whitworth be Surveyor of the Road that leads from Jeneto Road to a branch near Craddocks Plantation and that the two Wingo's and their Tythes John Chapman and James Hayes and their Tythes assist in Clearing the Same.

24 October 1751 O. S., Page 386
Ordered that Henry Farley be Surveyor of the Road that leads from Craddock's branch to the Church and that William Farley, William Jackson, Francis Jackson, John Brown, Benjamin Lockett, Daniel Farley and Samuel Farley and their Tythes Assist in Clearing the Same.

24 October 1751 O. S., Page 386
Ordered that John Burton Gent. be Surv^r. of the Road that leads from the Church Road to the Court House and that the Same hands as heretofore Assist in Clearing the Same

24 October 1751 O. S., Page 387
Ordered that Samuel Goode View the Way that leads from Saylors Creek Road to the Road that goes to Crawford's House and make report therefore at the next Court

Amelia County Order Book III

25 October 1751 O. S., Page 4
Ordered that the Bond Enter'd into by Thomas Anderson Samuel Tarry Clement Read and George Currey for the keeping up in good repair a Bridge over Appomattox River near Burtons be immediately Put in Execution So that Suit may be brought by the Persons impower'd by this Court to take the Same and the Expence thereof to born by the County and the Suit to be brought in this Court

25 October 1751 O. S., Page 4
Clement Read thereupon who is one of the Securities for Thomas Anderson on his bond for the building and keeping in repair a Bridge over Appomattox River near Burtons, moved the Court that they would direct and order the Commisioners by them appointed as Trustees for the County of Amelia to make an Assignment of the Bond by them taken for the Purpose aforesaid that he might put the Same in Suit for his indemnification it being Suggested to him and the Court that the Said Thomas Anderson intends Shortly to remove himself into the Province of Carolina, which the Court rejected they having already ordered a Suit on the Bond

25 October 1751 O. S., Page 4
Ordered that the Sheriff pay to John Nash Gent. twenty Six Pounds fifteen Shillings and Six Pence Current Money out of the Money Arising by the Sales of the Tobacco in his Hands for the building two bridgeges over Bush River

28 November 1751 O. S., Page 6
Grand Jury Presentment
The Surveyor of the Road from Nibbs Creek to Col^l. Cobbs Ordinary

28 November 1751 O. S., Page 8

William Booker Abraham Green & C^a }
Trustees for the County of Amelia Plts }
    vs } In Debt
Thomas Anderson, Samuel Tarry, Cement Reed }
& George Currie Defts }

[Margin note:] Continu'd for the resolves of Chesterfield Court

This Day Samuel Jordan came in to Court in the behalf of Thomas Anderson the Principle Deft. and undertook to Pay twenty Pounds Current Money as a proportionable Part Paid the Said Anderson by the Countyes of Henrico and Amelia for the building a Bridge over Appomattox River which Said Bridge was carried away by the inundation of the Said River after he had twice built the Same, This Court being Willing to receive the Said twenty Pounds and discharge the Said Anderson and his Securities provided that the Court of Chesterfield County will agree thereto; and for that Purpose Mr Richard Royal and Benjamin Harris are desired to acquaint the Court of Chesterfield with the Resolutions of this Court

28 November 1751 O. S., Page 10
Joseph Epperson haveing Built a Bridge over Appomattox River at Jennytoe the same was made use off the Sixth of this month

28 November 1751 O. S., Page 10
Daniel Coleman according to his Writing Obligatory having Errected and finished a Bridge over Appomattox River at or Near Burtons was made use of on the Second Day of October last.

12 February 1752 New Style, Page 21
George Walker Gent as Well for himself as many of the upper inhabitants of this County Presented a Petition for Clearing Appomattox River from Loggs and Trees and other incumbrances which may obstruct the Passing and repassing of Boats up and down the Same being read is ordered to be Certified to the General Assembly ---

27 February 1752 New Style, Page 27
In Pursuance to an Order from Chesterfield Court Produced here by Mr Benjamin Harris Which impowers M^r. Alexander Gordon to agree with Some Person to repair the Bridge over Appomattox River at Gordons which hath lately received Some damage and it being expedient the Same should be imediatley repaired this Court readily agreed to the Proportionable Part of the Charges according to the Purport of the Order here Produced from the Said Court of Chesterfield

28 May 1752 New Style, Page 43
Grand Jury Presentment
... the Surveyor of the Road from Richard Dennis above Woody Creek ...

23 July 1752 New Style, Page 59
Alexander Gordon produced an account of his Expences in Maintaining a bridge at Goods Amounting to twelve Pound five Shillings It is ordered that the Sheriff pay a Proportionable part thereof according to the Number of Tithables in this County ---

23 July 1752 New Style, Page 59
Ordered that the Sheriff pay to Thomas Belcher two Pound nineteen Shillings for building a Bridge over Sweat house Creek and keeping the Same in Repair Seven Years ...

23 July 1752 New Style, Page 59
Ordered that the Sheriff Pay to Daniel Coleman the Proportionable part of one hundred and twenty five Pound for building a Bridge over Appamattox River According to the Number of Tithables in this and Chesterfield Countys ---

27 August 1752 New Style, Page 64
Ordered the Sheriff pay John Nash Gent Ten pounds for Building a Brige over Bush River
Ordered the Sheriff pay Charles Anderson Seven pounds five Shillings for a Bridge over Bryery River

Ordered the Sheriff pay Thomas Morton Four pounds Ten shillings for a Bridge over Mountain Creek also Two pounds fifteen Shillings for a Repairing Buffilloe Bridge

27 August 1752 N. S., Page 64
Ordered the Sheriff pay Rich^d. Morton for Building a Bridge over Appomatox River between this and Cumberland our proportion of Seventy pounds according to Number of Tithables

27 August 1752 N. S., Page 66
Ordered the Sheriff pay Thomas Tabb Gent One hundred and Eleven pounds Sixteen Shillings and Eight pence being this Countys proportion of One hundred and Eighty pounds for Building a Bridge over Appomatox River at Jennytoe which bridge was built by Joseph Epperson and Company

23 November 1752 N. S., Page 73
Ordered the Sheriff pay Thomas Belcher Two pounds Nineteen Shillings for his Building a Bridge over the Swethouse Creek

23 November 1752 N. S., Page 74
Grand Jury Presentments
... The Surveyor of the Road from the Court house to George Elliotts

... the Survey of the Road from Mallorys Cree to Degernetts
The Surveyor of the road from the fork above Cheathams to the Widow Mayes. The Surveyor from Baldwins ordinary to Watsons race feild

25 November 1752 N. S., Page 79

To Charles Anderson Gent. for Reparing Bush & Bryery River Bridges	6..5
To Thomas Pain for reparing uper Flatt Creek Bridge	1..10
To Thomas Tabb Gent. as P his Account for reparing Jenny Toe Bridge & Prison	50..5..-3/4
To Duglass Puckett for reparing Sandy River Bridge and keeping the same in repare 7 years	11..5
To Samuel Cobbs for Burtons Bridge	1..10..0

* * *

To Duglass Puckett for reparing Bush River Bridge ---   8..

28 December 1752 N. S., Page 81
The Sherif by an Order of this Court having made Report that he had received of Thomas Anderson £20 Current Money whereupon the Court agree to discharge the said Anderson from his Bond Given to this County & Henrico for the Keeping a Bridg over Appomattox at Burton

23 March 1753 N. S., Page 90
Ordered that the Sheriff pay Alexander Roberts forty Shillings due to him for Building Genetoe Bridge

24 May 1753 N. S., Page 93
Grand Jury Presentments
... the Several Surveyors of the Road from the Head of Flatt Creek to head of Bush River
The Bridge over Flatt Creek near M^r. Tabbs is out of Repair
The Road from Wintocomake to Namaszeen is out of Repair
The Road from Rockey run to Wintocomake is out of Repair

24 May 1753 N. S., Page 93
Ordered that the Sheriff pay George More one Pound two Shillings & Six pence for Repairing the Bridge Over Mountain Creek.

24 May 1753 N. S., Page 93
Ordered that the Sheriff pay Thomas Morton Seven Pounds for Building a Bridge over Bush River and one over Bryor River.

25 May 1753 N. S., Page 97
Ordered that the Sheriff pay Ralph Shelton our Proportion of Six pounds for Building a Bridge over Nottoway River

26 July 1753 N. S., Page 115
Ordered that the Sheriff pay William Talley our proportion of seven pounds seventeen Shillings and six pence for building a Bridge over Namaszeen Creek and for keeping the same in Good repair for the Term or time of seven Years from the tenth day of March Last past

24 August 1753 N. S., Page 130
Ordered that if the Sheriff has any money in his hands, that he pay Alex. Roberts forty Shillings for repairing Knibbs Creek bridge

25 October 1753 N. S., Page 142
Daniel Coleman for the Consideration of 500 lb^s. of Tob^o. Agrees to build a bridge over Winticomake and keep the same in repair Seven years

26 October 1753 N. S., Pages 142-143
County Levy
To Thomas Dunnivant for Posts & Setting up 3 boards at the Cross Roads £ 1
To John Turner for 2405 Letters for boards on Roads   2405
To John Compton for puting up Boards and one Post   4 ---

* * *

To Daniel Coleman for a bridge which he is to build Over Winticomake Creek   500

22 November 1753 N. S., Page 146
Grand Jury Presentments
... the Surveyors from the head of Flatt Creek to Geo. Moor's, the Road from Colo. Cobbs Ord^y to the Court House, The Bridge Over Flatt Creek near James Fargusons out of Repair, No Sign Boards at the fork of the Roads near Henry Farley's the same in Cocks fork to Farleys, the same at the five forks, the Crose Road near George Hamms, Near Major Peter Jones, Near Goods Bridge the first fork above The Fork of the Road that Leads to Warwick at Col. Cobbs Ordinary, The Bridge Over Knibbs Creek near M^r. Andersons and Road to Col^o. Cobbs Ordinary out of Repair   The Sign boards Cross Roads at Col. Cobbs Ord^y. No Sign Boards above Bevills at Bottoms & d^o. forks below Bottoms No Sign boards at the fork where Tho Jones formerly lived

Index

This index is arranged by subject: Bridges and Causeways; Ferries and Fords; Land Features; Meeting Houses, Churches and Glebes; Mills and Mill Dams; Miscellaneous subjects; Other Counties and Courts; Personal Names; Race Grounds/Paths/Fields; Rivers, Runs, Springs, Creeks, and other Water Features; and Roads.

Bridges and Causeways

Anderson's bridge, 52,
Bridge built by James Anderson, 38
Mr. Anderson's bridge, 2
Mrs. Anderson's bridge, 16
Bridge below Mrs Anderson's quarter, 1
Andrew's bridge, 35
Appomattox River bridge.[see also River Bridge], 11, 13, 16, 19, 20, 25, 28, 31, 33, 41, 43, 50, 69, 70, 71
Appomattox River bridge at Bass's (near Genito), 20, 21, 22, 33
Appomattox River bridge (built by Thomas Anderson) at Bevills, 55, 69
Appomattox River bridge at Burton's, 23, 24, 29, 56, 68, 69, 70, 71
Appomattox River bridge between Fighting Creek and William Bass's quarter, 20
Appomattox River bridge at Genito (etc.) [see also Genito (etc.) bridge], 8, 9, 11, 36, 57, 69, 70
Appomattox river bridge at Goode's, 32
Appomattox River bridge at Gordon's, 69
Appomattox River bridge at Capt. Hudson's quarter, 15, 17
Appomattox River bridge at Liles Ford, 44
Appomattox River bridge at Sandy Ford, 57, 66
Appomattox River bridge at William Town's Plantation, 17
Appomattox River bridge at Webster's, 11
Appomattox River lower bridge, 21, 28, 32
Appomattox River upper bridge, 28, 56
Beaver Pond Branch bridge, 49, 52
Beaver Pond Creek bridge, 49
Bevill's bridge, 42, 55, 63
Richard Booker's Millstream bridge, 26
Boush River bridge [see Bush River bridge]

The Bridge, 10, 12, 14, 18, 24, 50
Bryary/Bryery River bridge, 54, 70, 71
Buffeloe/Buffilo/Buffilloe/Buffelow River bridge, 42, 43, 56, 59, 70
Burton's bridge, 17, 20, 46, 47, 49, 64, 71
Bush/Boush River bridge, 21, 24, 29, 34, 39, 48, 53, 54, 58, 65, 66, 68, 70, 71
Causeways, 6
Cellar/Celler Creek bridge, 3, 18, 49
Clement's bridge over Flatt Creek, 13
Combs's bridge, 12
Combs's bridge over Flatt Creek, 16, 30
Bridge over run near John Comb's, 33
Cradock's/Craddock's bridge, 14, 46
Deep Creek bridge, 2, 14, 18, 21, 34, 35, 45, 46, 49
Deep Creek bridge at Burton's, 49
Deep Creek bridge at Green's, 17
Deep Creek bridge at Peter Jones's quarter, 18, 47
Deep Creek lower bridge, 1, 11, 13, 49, 59, 62, 63
Farguson's bridge/causeway (causey), 61
Ferguson's bridge, 49
Ferguss's bridge/causeway, 55
Flatt creek bridge [see also Clement's bridge; Combs's bridge], 18, 23, 24, 30, 32, 33, 34, 35, 39, 41, 47, 50, 51, 52, 62, 63, 72
Flatt Creek bridge causeways, 39, 51
Flatt Creek bridge at Mr. Burtons, 17, 47, 64
Flatt Creek bridge at May's/Maye's, 46, 53
Flatt Creek bridge at Mr. Tabb's, 71
Flatt Creek lower bridge, 19, 35
Flatt Creek upper bridge, 23, 31, 32, 35, 40, 48, 71
Genito/Genetoe/Gennitoe/Jeneto/Jenito/Jennitoe/Jenny Toe bridge, 8, 9, 11, 13,

34, 36, 44, 45, 46, 48, 50, 53, 55, 57,
64, 69, 70, 71 34, 44, 45, 36, 57, 70, 71
Goode's/Good's bridge, 35, 37, 38, 44, 45,
48, 49, 59, 63, 70, 72
Great Nottoway bridge, 57, 60
Great Sailor's Creek bridge, 65
Harricane bridge, 60
Jeneto bridge [see Genito (etc.) bridge]
Jordan's/Jordon's bridge, 9, 14, 15, 60
Knibbs Creek bridge [see also Nibbs Creek bridge], 72
Knibbs Creek bridge at Col. Booker's/Mr Booker's, 17, 39
Leigh's bridge, 60
Little Creek bridge, 42
Little Nottoway bridge, 18, 41, 42, 46, 50, 58, 60
Little Roanoak bridge, 54
Mountain Creek bridge, 56, 59, 70, 71
Namaszeen Creek bridge   72
Namozain/Namazeen/Namoz.n/Namozain/Namozein/Namozn bridge, 1, 2, 3, 6, 7, 16, 29, 32, 34, 62, 72
Nibbs Creek bridge [see also Knibbs Creek bridge], 30, 63
Nottoway River bridge, 12, 20, 23, 31, 35, 36, 59, 60, 71
Nottoway River bridge at Dyer's Lick, 52
Nottoway River bridge at Stoker's, 59
Pole bridge, 36, 61
Bridge built by Col. Richd. Randolph, 10
River bridge [probably Appomattox River], 12, 16, 19, 21, 27
Upper River bridge   22
Sailors Creek bridge, 65
Sandy River bridge, 48, 53, 65, 71
(Old) Saponey ford bridge, 31
Sellar Creek bridge, 46
Sellers Creek bridge, 49
Smacks Creek bridge, 42
Smith's Creek bridge, 64
Stoker's bridge, 44, 59, 60
Stony bridge, 63, 64
Sweathouse Creek bridge, 67, 70
Webster's bridge, 11

West Creek bridge, 35, 40, 50, 53, 66
Wintercomake/Wintercomack/Wintercomeck/Wintereomake/Wintocomack/Wintocomeck/Wintycomeck/Witocomeck bridge, 9, 10, 40, 49, 53, 54, 58, 62, 66, 67, 72

Churches, Chapels and Glebes
Bush River Church, 43
Chapel, 15, 29, 31, 36, 37
Church, 2, 6, 7, 8, 9, 13, 15, 20, 27, 30, 33, 44, 47, 49, 55, 57 64, 66, 67, 68
Flatt Creek Church, 1, 30, 47
Namozain Church, 31
Nottoway Chapel, 2, 4, 9, 15, 25, 59
Nottoway Church, 55, 59
Nottoway Parish Upper Church, 56
Rocky Run Chapel, 18, 19, 30
Sandy River Chapel, 22, 51

Ferries and Fords

Appomattox River ferry boat, 2
Thomas Bevill's ferry/boat, 3, 5, 7, 10, 14
Crenshaw's ford, 36
Cut Banks ferry, 6
Hudson's ford, 34
Liles's/Lilis's/Lilles's/Lyles's ferry/ford, 2, 20, 21, 23, 44, 47, 52, 62
Namozain ferry boat, 7
Rattlesnake ford, 1
John Robertson's ford, 45
Rutledge's/Rutlidge's ford, 27, 34, 66
Sandy ford, 31, 36, 57, 66
Sappony ford, 1, 13
(Old) Saponey ford, 31
Vaughn's Creek old ford, 67

Land Features

Cut Banks, 2, 6, 7
Dyer's lick, 52

Hills on the other side of Flatt Creek, 14
Ridge between Amelia and Brunswick counties, 6
Ridge between Briery River and Buffalloe River, 27
Ridge between Little Nottoway and Great Nottoway, 50
Ridge of Nottoway, 16
Ridge between Nottoway and the Lazaretta, 6
Ridge between Peters Creek and Whetstone, 50
Ridge at the fork of Sandy Creek, 17
The Ridge, 21

Mills and Mill Dams

Abram Baker's mill, 58
Booker's mill, 7, 27, 45, 63
Mr. Booker's mill, 13
Richard Booker's mill, 13, 16, 19, 25, 26, 33
Clement's mill, 24, 28
Clement's old mill 32
William Clement's mill, 32
Mr. Cock's mill, 9
Richard Jones's mill, 23
Lovell's mill, 28
Maj. Munford's mill, 46
Nash's mill, 59
Col. Randolph's mill/Randolph's mill, 36, 67
Henry Robertson's mill, 26, 36
Henry Robinson's mill, 25
John Smith's mill, 32
James Walker's mill, 67
Watson's mill, 37

Miscellaneous Subjects

Akin's tenant 10
Anderson's quarter, 55
Henry Anderson's house, 3
James Anderson's house, 60
Mrs. Anderson's quarter, 1, 4
Thomas Anderson's house, 18
Atwood's plantation, 58
Baldwin's ordinary, 59, 71
William Bass's land/quarter, 20, 21, 22, 33
Bevill's line, 21
Maj. Bland's quarter, 31
Booker's Brick Yard, 35
Col. Booker's house, 39
Edward Booker's house, 7, 22
Edward Booker, Jr.'s land, 25
George Booker's house, 28
John Booker's house, 33
William Brown's house, 13
Browne's ordinary, 49
Buckskin, 1, 20
Buckweding, 33
Burton's line, 21
John Burton's house, 64
Carolina, 68
Hugh Chambers' plantation, 36
Chestnut Oak (on William Green's land), 34, 62
Cobbs's/Col. Cobbs's/Mr. Cobbs's ordinary, 35, 63, 69, 72
Samuel Cobbs's house, 42
Cock's quarter, 27
County line, 4, 35, 39, 40, 43, 50, 58, 59, 60
Court House, 2, 3, 7, 8, 10, 11, 15, 16, 17, 22, 25, 29, 32, 39, 44, 47, 59, 60, 61, 63, 64, 65, 66, 67, 68, 70, 72
Cox quarter, 39
Craddock's plantation, 68
Crafford's house, 15
Crawford's house, 38, 56, 62, 65, 66, 68
Dejarnett's smith shop, 58
The Folly, 64, 66
Geneto/Genetoe/Genito/Jenito/Jennitoe/Jenny Toe/Jennytoe [NOTE: may also refer to bridge or water feature], 8, 9, 11, 20, 34, 36, 44, 45, 50, 55, 57, 65, 68. 69, 70, 71
Abraham Green's house/quarter, 30, 42

William Green's land, 34
Stith Hardways's quarter, 55
Hawkin's plantation, 48
The Hatter's, 52
Capt. Hudson's plantation, 17
Capt. Hudson's quarter, 15
Charles Irby's house, 15
Jeneto (and alternate spellings) see Genito [NOTE: may also refer to bridge or water feature]
Abraham Jones's quarter, 1
Capt. Jones's quarter, 11
Peter Jones's/Capt. Peter Jones's quarter, 1, 18
Peter Jones's tenant, 25
Samuel Jones's quarter, 19
William Lee's quarter, 50
Letbetter's low grounds, 6
Samuel Major's plantation, 47
Alexander Marshall's quarter, 55
George Moore's house, 22, 37, 38, 52
William Mott's house, 5
Maj. Munford's land, 46
Capt. John Nash's quarter, 26, 38
Mr. Nash's house, 41
Mr. Nash's quarter, 14, 33
William Neale's house, 13
John Pride's plantation, 50
Randolph's lower quarters, 56
Col. Randolph's quarters, 5, 33, 39
Col. Peter Randolph's land, 41
Col. Richard Randolph's quarter, 6, 17, 26
Col. Richard Randolph's Mountain Creek Quarter, 27
Col. Richard Randolph's upper quarter, 37
Col. William Randolph's upper quarter, 34
William Ray's plantation, 51
Henry Robertson's house, 26
James Robertson's plantation, 41
Henry Robinson's house, 25
Benjamin Ruffin's quarter, 38
The School house, 34, 63
James Scott's house, 42
Mr. Sherwin's plantation, 19
Mr. Smith's fence/land, 34

Southall's ordinary, 35
Capt. Stark's quarter, 1
Capt. Stark's new quarter, 16
Capt. Starke's plantation, 12
George Steegall's house, 67
Thomas Tabb's house, 13
William Tinstall's house, 24
Towns's/Mr. Towns's quarter, 28, 42, 44
William Town's plantation, 17
Tunstall's quarter, 37
Upper Botton, 58
Edmund Walker's house, 63
Mr. Walker's plantation, 5
Mr. Walker's quarter, 59
Benjamin Ward's plantation, 38
Joseph Ward's house, 39
Ward's quarter, 14, 29
Warwick, 44, 48, 72
Watson's muster field, 21
Capt. Watson's Flatt Creek quarters and Mallary's Creek quarter, 49
White Oak on Flatt Creek, 4
Wilkinson's quarter, 29, 44
Williamson's land, 33
George Williamson's land, 21
Winterham, 64
John Worsham's quarter, 8, 55
Peter Wynne's ordinary, 46
William Yarborough's house, 29

Other Counties and Courts

Brunswick County, 6, 9
Brunswick County Court, 10, 23, 59
Brunswick County line, 11, 14
Chesterfield County, 70
Chesterfield County Court, 48, 55, 56, 69
Cumberland County, 70
Cumberland County Court, 55, 56, 57, 66
Goochland County, 20, 21, 33
Goochland County Court, 8, 9, 11, 15, 17, 21, 22, 33, 34, 38, 41, 44, 46
Henrico County Court, 2, 23, 24, 25, 29, 31, 32, 69, 71

Lunenburg County Court, 35, 52, 57
Lunenburg County line, 33, 35
Prince George County, 9
Prince George County Court, 2, 6, 32
Prince George County line, 15

Personal Names
[NOTE: Names may be spelled several different ways; all spelling variants should be checked; a blank (_____) indicates that the name is absent in the record]

_____
    Jack [slave], 35
    Will [slave], 35
Akin
    Hudson, 10
Alan
    Francis, 1
Alexander
    James, 31, 34
Allbright
    John, 49
Alsup
    Thomas, 49
Anderson
    _____, 1, 3, 7, 12, 13, 14, 16, 21, 39, 42, 52, 55, 59, 63, 64, 67
    Capt., 58
    Charles, 27, 41, 42, 43, 54, 56, 59, 70, 71
    Francis, 17, 18, 24, 32, 33, 39, 49, 57, 62
    Henry, 3, 7, 12, 16, 30, 33, 44
    James, 3, 6, 9, 16, 18, 23, 38
    James, Sr., 60
    John, 67
    Mr., 2, 14, 47, 51, 72
    Mrs., 1, 4, 16
    Paulen/Paulin/Pauling, 17, 32, 59
    Richard, 32
    Thomas, 9, 18, 25, 29, 32, 55, 59, 68, 69, 71
Andrews
    _____, 35, 63

Archer
    William, 20, 44, 46, 48, 53, 55, 57
Armstrong
    James, 67
Atcherson
    James, 54
Atkins
    Robert, 33, 54, 64
Atwood
    _____, 58
    James, 12, 37, 53
Austin
    Bat./Batho., 6, 17
Avery
    George, 14, 44
    John Allbright, 49
Bagley
    _____, 1
    George, 3, 16
Baker
    Abram, 58
    Caleb, 54
    Robert, 31
    Saml., 31, 34
Baldwin
    _____, 62, 71
    John, 43
    Will, Jr., 22
    William, 48, 58
Baldwinn
    _____, 59
Ball
    James, 27
Barnes
    William/Wm., 19, 29, 55, 64
Barns
    William, 10
Bass
    _____, 21
    Christopher, 54
    William/Willm., 20, 21, 22, 33
    William, Jr., 54
Bates
    _____, 44

Battes
　_____, 43
Batts
　_____, 39
　Henry, 20, 58
　William, 20, 58
Bayley
　George, 23
Beasly
　Stephen, 33
Beisley
　_____, 37
　Ambros, 36
　Richd./Rd., 8, 11, 19, 37
　William, 8
Belcher
　John, 36
　Thomas, 47, 53, 70
　Will./William, 16, 36, 48
Bell
　Zecheriah, 28
Bennett
　James, 66
　Richard, 66
Bennitt
　Benja., 19
Benson
　_____, 12
　John, 2, 13, 14, 15, 31
Bentley
　John, 44
　Stephen, 29
Bevil
　Thomas/Thos., 5, 31
Bevill
　_____, 21, 42, 55, 63, 72
　Danl., 21
　Essex, 21
　Thomas/Thos., 3, 7, 10, 14, 21, 24, 63
Bibb
　John, 27
Bibbs
　John, 54
Blake
　James, 67

Blake
　Robert, 67
Blanchet
　John, 48
Blanchett
　_____, 12
　John, 8, 22, 42
Bland
　Maj., 31
　Mr., 18
Bolling
　Alexander, 58
Booker
　_____, 2, 3, 7, 21, 22, 27, 34,
　　42, 45, 63
　Col., 35, 38, 39
　Edmund/Edmd., 16, 26, 44
　Edmund/Edmd., Jr., 34, 38, 44, 46, 62
　Edward, 2, 7, 11, 15, 16, 17, 20, 22,
　　32, 34, 41
　Edward, Jr., 25, 37, 44
　George, 27, 28, 33, 39, 51, 61
　John, 33, 34, 35, 40
　Maj., 45, 51
　Mr., 12, 13, 16, 17, 38, 55
　Richard/Richrd., 3, 8, 9, 11, 13, 16, 17,
　　19, 20, 21, 26, 33, 38, 40, 41, 45, 46,
　　47, 48, 56, 63, 66
　Richard, Jr., 45, 63
　Richard, Maj., 25
　Thomas, Jr., 67
　William, 11, 17, 23, 24, 25, 29, 32,
　　38, 41, 45, 46, 55, 69
Booth
　Thomas, 54
　Thomas, Jr., 29, 48
Boram
　_____, 63
　Richard, 16, 35, 48
Boston
　Hugh, 7, 12
Bottom
　_____, 72
　Thomas, 3, 22, 26, 47

Botts
    Mr., 21
Bowman
    Robert, 11
Bowry
    Thomas, 31
Braggs
    John, 27
Branch
    _____, 12
    Mr., 33
Branson
    William, 37
Branton
    William, 37
Brathwait
    _____, 49
    Edward, 49
Brathwett
    Edward, 34, 56
Bridgford
    _____, 12
Bridgforth
    John, 54
Brooks
    _____, 36
    Thomas, 16, 22
    William, 38
Broomfeild
    James, 27
    James, Jr., 27
Brown
    _____, 12
    Daniel, 11, 30, 32
    John, 68
    Laz., 8
    William, 9, 13, 36, 40, 41
Browne
    _____, 49
    William, 52
Bruce
    Alexander, 55, 59
Brumfeild
    Wm., 15

Bullington
    Benja., 49
Burk
    Charles, 7
    Chas., Jr., 12
Burke
    Charles, 39
    George, 39
Burkes
    Charles, 3
Burks
    Charles, 30, 64
    Charles, Sr., 32
Burton
    _____, 8, 10, 15, 16, 17, 20, 21, 29,
        46, 47, 49, 56, 61, 64, 68, 69, 71
    Abraham, 1, 3, 21, 23, 24, 28
    John, 5, 7, 8, 10, 13, 17, 24, 46, 64,
        65, 68
    Mr., 17, 38
    Thomas/Thos., 12, 20, 21, 47
Cabaness
    Mathew, 9
Cabanis
    Mathew, 25, 30
Cabiness
    Matthew, 45, 47
Caldwell
    George, 34
    John, 67
Callaway
    _____, 65
Callicoat
    James, 50
Callicott
    William, 22
Callicutt
    William, 42
Cambell
    Rober, 34
Canister
    John, 19
Cannon
    Simcock, 58

79

Carrell
    John, 37
Carter
    Theodorick/Theo., 34, 43
    Thomas, 49
Cates
    Curtis, 26
Causton
    Charles, 65
Certain
    Thomas, 38
Chambers
    _____, 36
    Hugh, 21, 33, 36
Chandler
    Isaac, 58
    Thomas, 50
Chapman
    John, 68
Cheatham
    _____, 19, 44, 45, 62, 71
    Charles, 48
    James, 27, 44, 46, 59, 62
    Jonathan, 53
    Leonard, 65
Chessright
    John, 43
Childress
    Phillamon, 15
Childry
    Jeremiah, 22
    John, 19, 25
    Widow, 25
Childs
    Henry, 67
    John, 67
Chisham
    John, 36
Chissum
    John, 61
Chistam
    John, 51
Clark
    Henry, 40, 44
    John, 26
    Richard, 25
Clarke
    _____, 2
    John, 49, 59
    Peter, 49
    William/Wm., 1, 5, 7, 16, 33
    Witt, 21
Clay
    Charles, 45, 53
    John, 48, 54, 62, 67
Clayborn
    _____, 54
    Leonard, 59
Clement
    _____, 13
    Mr., 61
    William, 6, 32, 33
    William, Jr., 32
Clements
    _____, 24, 28, 32
    Ben, 32
    Mr., 32
    William, 24
Clemment
    John, 36
    William, 5
Cobb
    Mrs., 21, 34
Cobbs
    _____, 63
    Col., 35, 69, 72
    Mr., 38, 51, 63
    Mrs., 29
    Samuel/Saml., 2, 11, 16, 20, 40, 42, 45, 48, 55, 64, 71
Cock
    _____, 15, 20, 27, 36, 57, 72
    Abraham, 9, 20, 23, 35, 57, 59
    Mr., 9, 43, 60
Cocke
    Abraham, 42, 46, 52, 58
    Mr., 40
Coffee
    Peter, 67

Cole
    John, 32
Coleman
    Daniel, 6, 11, 13, 16, 18, 29, 40, 49, 66, 67, 69, 70, 72
    Robert, 11
Coles
    _____, 53, 54
    James, 29
Coleson
    James, 51
Collins
    _____, 11
    James, 6, 12
    Stephen, 26, 27, 30, 33
Combs
    _____, 12, 16, 30
    John, 16, 33, 35, 38
Compton
    John, 51, 72
Connolly
    Charles, 40
Conway
    _____, 49
Coock
    George, 67
Cotteril
    Charles, 21
Cottrell
    Charles, 14, 29
Cousens
    Robert, 34
Covington
    Edmund, 22
    Edward, 19
    Thomas, 3, 4, 10, 14, 45
    William, 23
Cox
    _____, 39
    Edward, 20, 31
    John, 11, 54
Cozins
    Robert, 62
Craddock
    _____, 4, 46, 68
    William, 30
Cradock
    _____, 8, 14
    Wm., 19
Crafford
    _____, 15, 51
Craford
    _____, 11
Crawford
    _____, 8, 38, 56, 62, 65, 66, 68
Crawley
    Mrs., 33
    William, 2, 54
Crenshaw
    _____, 36
    William, 14
Cross
    William, 20, 58, 60
Crowder
    James, 31
Cruchfield
    Richard, 50
Crump
    Stephen, 28
Cryer
    William, 43
Cumpton
    John, 62
Cunningham
    Alexander, 43
    James, 31
    John, 31
Currey
    George, 68
Currie
    George, 29, 69
Dabney
    _____, 2, 5
Dandy
    _____, 3, 16, 26, 31, 56, 60
    William, 26, 56
Daniel
    James, 67
Davidson
    John, 47, 66

Davis
  _____, 33
  George, 54
  John, 11, 27
  Peter, 26, 39, 64
Dawson
  _____, 19, 51
  Henry, 17, 28
  John, 4, 5, 52
  Mrs., 17
Degarnett
  Daniel, 39, 42
Degernett
  _____, 50, 71
Dejarnett
  _____, 58
  Daniel, 21, 54, 58
Dejarnette
  Daniel, 56, 61
Dennis
  Nathl., 8
  Richard, 26, 70
Dicke
  Henry, 50
Dicks
  James, 8
Dix
  James, 19
Dodson
  Greenham, 44, 60
Doss
  James, 50, 66
Dow
  Andrew, 67
Dunafant
  William, 24, 28
Dunifant
  William, 16
Dunnavan
  Thomas, 45
Dunnivant
  Thomas, 72
Dupuy
  Peter, Jr., 50

Dyer
  _____, 52
  John, 37
Eastis
  Abraham, 53
  Ambrose, 58
Echols
  John, 28
  Richard, 65
  William, 4
Eckhols
  William/Will., 6, 17, 32
Edwards
  Samuel, 48
  William, 27
Elkin
  Ralph, 27
Ellin
  Samuel, 59
Ellington
  David, 26
  John, 33, 36
Elliot
  John, 57
Elliott
  George, 57, 63, 65, 70
Ellis
  John, 8, 14, 22, 25, 56
  Richard, 56, 60
  Thomas, 14
Epperson
  Joseph, 57, 69, 70
Estis
  Elisha, Jr., 55
  Mr., 55
Evan
  Saml., 31
Evans
  George, 33, 39, 41, 55
  Giles, 67
  John, 14
  Robert, 14
  William, 14, 15, 59
Ewen
  James, 31

Ewing
    George, 34, 67
    James, 67
    Thomas, 67
    William, 53, 67
Fannell
    Bryant, 43
Farguson
    _____, 61
    James, 72
    John, 33, 35
    Robert, 33, 60
Farley
    _____, 64, 72
    Daniel, 68
    Henry, 68, 72
    Herb., 13
    John, 57
    Lawrence, 22
    Samuel, 68
    Stewart, 57
    William, 34, 68
    Wm., Jr., 15, 27
Farloe
    Henry, 10
Farlow
    _____, 65
Fenell
    Bryan, 2
Fenning
    Bryan, 6
Fergerson
    John, 20
Ferguson
    _____, 49
    James, 64
    John, 19
    Robert, 41, 51
    William, 18
Fenney
    Bryan, 4
Ferguss
    _____, 55
Fergusson
    James, 55

    Robert, 52
    Robert, Sr., . 55
    William, 51, 55
Field
    Robert, 33
Fisher
    _____, 14
Fleming
    Robert, 48
Flournoy
    _____, 53
    David, 61, 64
    Mat, 61
Flouronoy
    _____, 39
Flyn
    Laugh./Laughland, 12, 20
Forbush
    Robert, 67
Ford
    Frederick/Fredk., 18, 32
    Hezekiah/Hez., 11, 13, 18, 24, 46, 51,
        52, 56, 61
    Mr., 32, 44, 59
Forguson
    John, 1, 3, 16
    Robert, 12, 15, 16
    Will, 16
Forrest
    Abraham, 56
Forster
    George, 40, 42, 47, 51
    John, 31, 51
    Thomas, 7, 8, 12
Foster
    _____, 62
    George, 17, 37, 49, 61
    James, 56
    Thomas, 31, 51
    William, 45, 59, 62
Franklin
    _____, 5
    Edwin, 3
Frazier
    Alexander, 29, 66

Friend
    Edward, 44
Fulton
    Thomas, 67
Furgusson
    Peleg, 58
Gaines
    Henry, 36
Galaspie
    Patrick, 54
Galespy
    Robt, 34
Gaulden
    John, 34
Gentry
    John, 38, 65
Gibbs
    John, 38
Gibson
    Noel, 50
    Valentin, 66
Gillenten
    _____, 47
Gillingtine
    _____, 27
Gillington
    John, 33
Gillintine
    _____, 32
    John, 24
    Nicholas, 36
Good
    _____, 35, 37, 38, 44, 45, 70, 72
Goode
    _____, 32, 48, 49, 59, 63
    Samuel, 64, 68
Goodwin
    George, 36
Gordon
    _____, 69
    Alexander, 69, 70
    Samuel, 37
Grainger
    John, 25
    Joseph, 15

Granger
    _____, 12
Gravely
    James, 30
Gray
    Alexander, 23
Green
    _____, 17, 62
    Abraham/Abram, 2, 17, 21, 23, 24, 25,
        29, 30, 32, 34, 35, 42, 48, 54, 55, 56,
        58, 66, 67, 69
    Francis, 52
    Mr., 49, 52
    Thomas, 31
    Thomas, Jr., 63
    Thomas, Sr., 62
    William, 1, 4, 34
Gresham
    Robert, 67
Griffin
    _____, 30, 47, 60
    Anthony, 21, 22, 26, 37, 39, 40, 41, 42,
        61, 65
Gross
    Edmund, 21, 27
Guilintine
    _____, 61
    Nicholas, 61
Guillington
    _____, 62
    Nicholas, 40
Hall
    _____, 3, 26
    James, 25
    John, 12, 23
    William, 67
Ham
    George, 36
Hamm
    George, 6, 36, 72
Hammack
    Bennedick, 44
Hammond
    Lewis, 40, 43

84

Hamond
  Duncomb, 5
Hampton
  John, 6
Hannah
  John, 67
Hardaway
  Mr., 29
  Stith, 40, 47
Hardcastle
  William, 26
Harden
  _____, 10, 11
  John, 47
Hardin
  John, 14, 33, 49
Harding
  _____, 26
Hardway
  Stith, 55
Hardy
  Thomas, 32
Harper
  Edward/Edwd., 8, 19, 62
  Solomon, 4
Harris
  Benjamin, 56, 69
  John, 15, 27
Harrison
  Col., 16
Hart
  Abraham, 50
Haskins
  Edward, 14, 21
  Thomas, 53, 65
Hatcher
  William, 12
Hatchett
  Wm., 21
Hawkins
  _____, 48
  Benjamin, 56, 63
  Thomas, 56
Hawks
  Abraham, 19

  Josiah, 19
Hayes
  James, 68
  John, 19, 26, 36
Haynes
  Capt., 43
Hendrake
  Benjamin, 57
  Hantz, 66
Hendrick
  Benjamin, 27
  Hance, 15, 18, 27, 33
  Hance, Jr., 15
  John, 15, 27
Hightower
  John, 43
  William, 43
Hill
  _____, 17
  George, 40, 44
  James, 44
  Jno., 36
  William, 5, 67
Hilsman
  Mathew, 36
Hinton
  Christopher/Christo., 1, 18, 19
Hix
  Richard, 7, 11, 26
Holderness
  Robert, 53
  Samuel, 53
Holloway
  John, 37
House
  Thomas, 50
Howell
  John, 38
Hubbard
  Benjamin, 55
Hudgins
  John, 35
Hudson
  _____, 26, 33, 34, 35, 36, 40, 67
  Capt., 15, 17

Hudson (continued)
    Isaac, 2, 4
    James, 23
    John, 11, 27
    Josiah, 30
    Samuel, 4, 5
    Thomas, 32
    William, 5
Hughes
    John, 44
Hulm
    William, 23, 25
Hurt
    _____, 19
    Abraham, 52
    John/Jno., 4, 32, 59, 62
    Moses, 58
Hutcherson
    Charles, 55
    William, 34
Hutchinson
    William, 51
Irby
    _____, 20, 22, 25, 26, 47
    Capt., 29, 35
    Charles, 5, 9, 12, 13, 15, 23, 25,
        26, 35, 37, 42, 46, 50, 52, 55, 57
    Mr., 1, 40, 41
    Peter, 33
Isbell
    Henry, 24
Israel
    Judah, 22
Jackson
    _____, 12, 20, 40, 60
    Charles, 43
    Edward, 43
    Francis, 68
    James, 36, 43
    John, 2, 4, 6, 12
    Mark, 57
    Matthew, 22, 42
    Thomas, 4, 6, 57
    William/Wm., 6, 11, 13, 15, 28, 34,
        39, 43, 44, 68

Jefferson
    _____, 16
    Field, 40
Jenkins
    John, 8
    Joseph, 8
Jennings
    Elkanah, 67
    Robert, 67
Johnson
    Charles, 51
    Henry, 21
    John, 36
    Sill, 21
Jones
    Abraham/Abra., 1, 18, 19, 51
    Abraham, Jr., 46
    Capt., 11, 53
    Daniel, 58, 59
    Edward, 29, 43, 47, 61
    George, 37
    John, 16, 43
    Maj., 47
    Peter, 18, 25, 46, 47
    Peter, Capt., 1, 2
    Peter, Maj. 43, 72
    Richard, 6, 10, 14, 46, 64
    Richard, Capt., 2
    Richard, Jr., 67
    Richard, Maj., 16, 23
    Saml., 19
    Thomas, 9, 11, 26, 59, 72
    Thomas William, 58
    Tobitha, 43
    William, 27
    Wood, 14, 32, 35, 40, 54, 55, 56, 67
Jordan
    _____, 9, 12, 14, 36, 58, 60
    Samuel, 4, 9, 12, 69
Jordon
    _____, 15
    Saml., 15
Keatly
    William, 4

Keeling
    Osborn, 30
Kelley
    William, 67
Kennon
    _____, 50
    William/Wm., 38, 41
King
    William, 65
Lawson
    Mr., 57
Lax
    James, 66
Leath
    Arthur, 60
Lee
    William, 50
Leigh
    _____, 60
    Arthur, 5, 8, 9, 15, 20
Leister
    Henry, 41
LeNeve
    _____, 16
Leonard
    John, 12
    Thomas, 12
Lester
    _____, 36
Letbetter
    _____,
Leverett
    _____, 8
    John, 6, 9, 11
Lewelling
    Daniel, 39
Lewis
    William, 53
Leyton
    Hugh, 8, 19
Liggon
    Henry   19, 53, 65
    William, Jr.   57
    Willm.   51

Ligon
    Henry, 29, 39
    Joseph, 30
    William, 29, 40
Liles
    _____, 23, 44, 47
Lilles
    _____, 52
Lilis
    _____, 21
Lisles
    _____, 62, 63
LittleJohn
    Joseph, 31
Lockett
    Benjamin, 68
Lorton
    Thomas, 28
Loudone
    Hugh, 49
Lovell
    _____, 28
Loving
    Richard, 17, 31
Lyan
    Elisha, 50
Lyles
    _____, 6, 7, 15, 20, 23, 51
    David, 1, 2, 6
Lyon
    Elisha, 66
Mackadoo
    Andrew, 67
Mackbride
    Manaseh, 67
Mackew
    James, 67
Mackfield
    Manase, 67
Macklew
    William, 67
Magehee
    Jacob, 14
Magehen
    Jacob, 11

Major
    John, 36
    Samuel, 36, 47, 49, 59, 62
Man
    Catlet, 55
    John, 54
Manear
    William, 48
Manire
    William, 40, 43
Mann
    _____, 33
    Cattlin, 61
    Francis, 21, 36
    John, 28
    Robert, 21
    Saml., 21
Markam
    Thomas, 34
Markham
    Thomas, 20, 23
Marshall
    Alexander, 55
    William/Will./Wm, 4, 7, 12, 21
Martin
    John, 9, 27, 52, 58
    John Robert, 31
    Robert, 67
Matthews
    Samuel, 67
Maulden
    John, 31
Mauldin
    John, 51
May
    _____, 46
    John, 50, 66
    William, 60
Mayes
    _____, 26, 53
    Gardiner/Gardner, 53, 56, 62
    John, 8, 26
    Mathew, 25
    Widow, 71
    William, 8, 14, 22, 25, 56, 57, 58, 62

Maynard
    William, 58
Mays
    William, 5
McDearman
    Michl., 21
Mcgehee
    Jacob, 27
Meadows
    Joel, 17, 51
Meak
    Guy, 12
Meredith
    Mr., 29, 44
Miller
    William, 67
Mitchel
    Mrs., 56
Mitchell
    James, 39
    John, 14
    Thomas, 19
    Walter, 55, 66
Mole
    William, 7
Moody
    Robert, 11, 55
Moor
    George, 27, 72
    James, 29
Moore
    George, 22, 33, 37, 38, 39, 41, 49, 52, 54, 58, 59, 61, 64
    William, 47, 48
More
    George, 71
Moreau
    Lawrence, 67
Morgan
    Joshua, 43
    Saml., 21
Morris
    Isaac, 57
    John, 40
    John, Jr., 51

Morrow
    John, 67
Morton
    John, 58
    Joseph, 5, 11, 27, 42, 54, 56, 58
    Joseph, Jr., 6, 34, 43
    Joseph, Sr., 10, 14
    Richd., 34, 70
    Thomas, 11, 30, 36, 70, 71
Motley
    Joseph, 21, 22
Mott
    William/Will., 5, 11
Mullins
    John, 14, 21, 27
Mullord
    Jos., 8
Munford
    Maj., 18, 19, 46
Murray
    _____, 63
Mutlee (Mutloe?)
    Joseph, 19
Nance
    John, 1, 4
    John, Jr., 5
    John, Sr., 5
Nash
    _____, 9
    John, 21, 24, 38, 41, 42, 48, 53,
        57, 58, 65, 68, 70
    John, Capt., 26
    Mr., 9, 10, 11, 14, 21, 29, 33, 38, 39,
        41, 43, 51, 65
    Mrs., 34
Neal
    _____, 36
    Roger, 21
Neale
    David, 12, 21
    Roger, 12
    Stephen, 12, 21
    William, 13
Neil
    Arthur, 67

Nelson
    Henry, 27
    Thomas, 27
Nicks
    Benjamin, 66
    Edward, 66
    James, 65, 66
    Valentine, 66
    William, 66
Nix
    Edward, 50
    James, 50
Nixon
    Hugh, 27, 54
Olive
    James, 25
Oliver
    James, 47, 55, 61
Ornsby
    Rev. Mr. John, 26,
    the Parson, 37
Osborn
    _____, 14
    Edwd., 21
    John/Jno., 12, 16, 21, 29
    Thomas, 20
    William, 33
Owens
    John, 37
Pain
    John, 54, 60
    Thomas, 71
Palmer
    Ellis, 58
Parks
    James, 31
(the) Parson [see Rev. Ornsby]
Parush
    William, 4
Payne
    John, 50
    Josiah, 50
Perdue
    Jno., 21

Perkerson
    Ralph, 21
Pettis
    Thomas, 35, 48, 63
Pettiss
    Thomas, 35
Phips
    John, 67
Pigg
    Paul, 6
Pincham
    Samuel/Saml., 3, 13, 16
Pitchford
    Saml., 21
Pitman
    John, 31
    William Jr., 31
Pledger
    Phil/Philip/Phillip, 36, 56, 60
Pollard
    George, 17
    Joseph, 51
Pool
    William, 4
Popham
    John, 53
Porter
    Thomas, 39
Potter
    _____, 36
Powell
    Hez., 6
Poythress
    Elizabeth, 43
Pride
    John, 7, 21, 30, 40, 50
    Mr., 33
Prisnall
    James, 50
Pucket
    Duglas, 28
Puckett
    DouglasDougls./Duglas, 21, 29, 39, 53, 66, 71
Quin

William, 50
Ragsdale
    George, 42
    John, 5, 43
Randolph
    _____, 26, 53, 56, 64, 65, 67
    Col., 5, 11, 33, 36, 39
    Peter, Col., 41
    Richard, Col., 6, 10, 17, 26, 27, 31, 37
    William, Col., 31, 34, 52
Ray
    William, 50, 51
Read
    Clement, 23, 29, 68
    Mr., 21
Reams
    Edward, 34
    Thomas, 21
Red
    William, 58
Reed
    Cement, 69
    Clement, 41
Rice
    Francis/Franc., 8, 19, 27
    Joseph, 37
    Michael, 37
Rickey
    Alexander, 27
    Charles, 27
    Hugh, 27
Ritchey
    John, 54
Roberts
    Alexander, 45, 50, 71, 72
    John, 24, 37, 64
    William, 66
Robertson
    Christo./Christopher, 1, 3
    Christo., Jr, 7
    Edward, 7, 18, 24, 25, 45
    Henry, 26, 36
    James, 38, 41
    John, 38, 41, 45
    Nathaniel, 25, 47, 60

Robinson
    Henry, 25
Row
    Michl. McDearman, 2
Rowland
    Robert/Rob., 7, 18, 22, 45
Rowlett
    Peter, 33
Royal
    Richard, 69
Ruffin
    Benjamin, 38, 65
Russell
    William, 27
Rutledg
    _____, 9
    Richd., 21
Rutledge
    _____, 34, 66
    James, 19
Rutlidg
    James, 21
    Richd., 21
Rutlidge
    _____, 27
    James, 29
Sadler
    Wm., 51
Sawney
    _____, 53, 67
Scott
    Doctor, 44
    James, 42
    Joseph, 8, 9, 11, 20, 21, 22, 31, 32,
        33, 34, 37, 38, 45
    Mr., 44
    Mrs., 50
Seay
    Gesse, 59
    Jacob, 12, 24, 36
    Jesse, 32
    Joseph, 31
Seircey
    William, 30
Shannon
    William/Wm., 4, 8
Shelton
    Joseph, 34, 43
    Ralph, 50, 71
Sherwin
    Mr., 19
Shinnaison
    John, 66
Shorts
    William, 26
Silcock
    William, 27
Simmons
    Charles, 67
    John, 44
Siresey
    _____, 36
Smith
    _____, 64
    James Mitchell, 45
    John, 17, 32, 56, 62
    John, Jr., 31
    John, Sr., 31
    Mr., 34
    Thornton, 45, 48, 53
    Widow, 56
    William, 67
Southal
    William, 51, 55
Southall
    _____, 35, 47
    Dacey, 38, 57
Spellers
    Francis, 30
Spencer
    _____, 29
    Mr., 44
    Thomas, 44
Spiner
    _____, 16
Spinner
    _____, 16, 20, 62
Spradling
    Charles, 38
    John, 65

Stadey
    Robert, 44
Stanley
    William, 23
Stark
    Capt., 1, 16
Starke
    Capt., 12
Steegall
    George, 67
Stegall
    Geo., 33
    John, 31
Steward
    George, 27
Stewart
    George, 33, 39
Stewert
    _____, 12
Stock
    _____, 62
Stocks
    _____, 4, 10, 17, 18, 27, 31,
        32, 33, 49, 51
Stoker
    _____, 44, 46, 59, 60
    Robert, 12, 44, 54
Stokes
    _____, 59
    Mr., 65
Stone
    Richard, 60
    Thomas, 67
    William, 40, 43, 48
Sunderland
    Samuel, 36
Tabb
    Capt., 12
    Mr., 37, 71
    Thomas, 9, 11, 13, 20, 21, 22, 23, 24,
        25, 29, 31, 32, 34, 36, 38, 44, 53, 55,
        56, 57, 65, 66, 70, 71
Talbert
    Math., 6
Talbot
    Mathew, 6
Talbott
    Mathew, 7, 23
Talley
    Abraham, 34
    John, 29, 45
    William, 72
Tally
    _____, 29
    Henry, 4
Tanner
    _____, 4, 36, 59, 67
    Lewis, 1, 3
    Lodwick, 12, 33, 46
    Mr., 16
Tarry
    Mr., 45, 49, 52
    Samuel/Saml., 19, 29, 30, 31, 36, 68,
        69
Tatum
    Josiah, 21
Taylor
    _____, 12
    Edwd., 55
    John, 37, 59
    Robert, 2, 22, 40, 43, 52
    Thos., 9
Thaxton
    James, 54
Thomas
    _____, 18, 19, 20, 60
    John, 9, 15, 20, 50, 60
    Mr., 8
Thompson
    John, 31
    Peter, 16
    Robert, 16, 33, 36
    Roger, 40
Thomson
    John, 22
    Peter, 21
    Robert, 21
    Roger, 35
Thurman
    Richard, 66

Thurman (continued)
    William, 66
Thweat
    Edward, 23, 25
Tinstall
    William, 24
Townes
    _____, 28
    William, 28
Towns
    _____, 36
    John, 17, 37, 50, 54, 64, 65
    Mr., 16, 42, 44, 51
    William, 17, 50, 65
Tromer
    John, 45
Tucker
    Daniel, 53, 62
    George, 28
    Robert, 29, 62
    Robert, Jr., 1
Tukaer
    _____, 50
Tunstall
    _____, 37
    Mr., 19
Turner
    John, 7, 10, 14, 22, 24, 72
Turpin
    Thomas, 37
Twitty
    John, 54
Vaughan
    _____, 8, 36, 65, 67
    Abraham, 38
    Isham, 25
    Lewis, 22, 25
    Robert, 3, 5, 8, 11, 22, 25, 30, 43
Vaughn
    _____, 17
    Robert. 30
Waddall
    John, 37
Wade
    Hampton, 56, 59

Walberton
    Thomas, 49
Walhall
    Henry, 14
Walker
    _____, 9, 21
    Capt., 37
    Edmund, 61, 63
    George/Geoge., 5, 9, 19, 48, 53, 56, 57, 65, 69
    James, 65, 67
    Mr., 5, 8, 9, 29, 43, 50, 59
    Thomas, 22
    Wm., 22
Wallis
    Saml., 34
Walters
    _____, 53
    Henry, 19
    Thomas, 42
Walthall
    Christopher, 46, 63
    Richard, 12
Ward
    _____, 14, 26, 29
    Benjamin, 38, 44
    Henry, 31, 55
    Joseph, 39, 40
Warren
    _____, 16
Washbun
    _____, 61
Watkins
    _____, 33
    Joel, 53
    John, 34
Watson
    _____, 20, 21, 25, 37, 47, 59, 61, 62, 71
    Capt., 11, 21, 32, 42, 47, 49, 54
    Mr., 45, 47
    William, 8, 19, 23, 25, 35, 40, 48, 49
Watsons
    Capt., 21

Wawmock
    Abraham, 53, 65
    Abraham, Jr., 65
    Thomas, 65
    William, 65
Webster
    _____, 11, 67
    Peter, 45, 63
    Thomas, 33
Wells
    Barnaby, 19
West
    _____, 29
    Ephrim, 26
    John, 26
Westbrook
    Charles, 26
    William, 26
White
    George, 56, 58
Whitton
    Richard, 57
Whitworth
    _____, 29
    Thomas, 55, 64, 68
    Thos., Jr, 29
Wilkinson
    _____, 14, 29, 44
    Martin, 58, 61
    William, 15, 45
Willard
    John, 19
Williams
    Bellington, 23
    Thomas, 26
Williamson
    _____, 33
    George, 20, 21, 22
    Thomas, 39, 51
    William, 56
Willis
    Maj., 37
Willison
    Danl., 22
Willson
    John, 22
Wilson
    Daniel, 42
Wimbush
    James, 54
    Mr., 58
Wingo
    _____, 68
Winingham
    _____, 35
    John, 2
Winn
    John, 14
Witt
    Richard, 37
Womack
    Matthew, 30
    Richard, 27
    William, 38
Womawk
    _____, 51
Woodson
    Benjamin, 66
    Richard, 27, 34, 42, 43
Worsham
    _____, 36
    Capt., 16, 42
    Daniel, 42, 63
    Essex, 38
    John, 8, 55
Wynn
    Peter, 44, 60
Wynne
    Peter, 31, 44, 46
Yarborough
    Samuel, 49
Yarborrough
    _____, 50
Yarbro
    Henry, 7
    William, 6
Yarbrough
    _____, 41, 45
    Hen., 36
    Jos., 37

Yarbrough (continued)
    Thomas, Jr., 36, 41
    Thos., 36
    William, 29
Yarbrow
    _____, 20
    William, 16

Race Grounds/Paths/Fields

Dandy's race paths, 3, 16, 26, 31
Dawson's race path, 51
Hudson's race paths, 36, 67
Race paths near Mr. Irby's, 1
Neal's race paths, 36
Race paths, 3, 20, 29, 61
Watson's race ground/path/field, 25, 59, 61, 62, 71
Andrew Wawmock's race paths, 53

Rivers, Runs, Springs, Creeks, and other Water Features

Andrews Branch, 63
Appamatox/Appamattox/Appomatox/ Appomattox River [see also the (Appomatox?) River], 2, 8, 9, 11, 13, 15, 16, 17, 19, 20, 21, 22, 23, 24, 25, 27, 28, 29, 31, 31, 32, 33, 36, 41, 43, 44, 50, 55, 56, 57, 65, 66, 68, 69, 70, 71
    Head of, 65
Beaver Pond Branch, 1, 49, 52
Beaver Pond Creek, 49
Bent Creek, 3
Birchen Swamp, 4
Booker's Brick-Yard Branch, 35
Richard Booker's Millstream, 26
Boush River [see Bush River]
Bryery/Briery/Bryary/Bryor River 27, 54, 58, 59, 64, 70, 71
    Head of 59
Buckskin, 1, 20

Buffalo/Buffeloe/Buffelow/Buffilloe/Buffilo River, 5, 27, 34, 42, 43, 49, 56, 59, 70
    Fork of, 49
    Mouth of, 27
Buffelow-Bed Creek, head of 60
Bush [Boush] River, 4, 5, 9, 14, 19, 20, 21, 22, 24, 25, 26, 27, 28, 29, 34, 35, 36, 37, 39, 41, 43, 44, 47, 48, 51, 53, 54, 56, 57, 58, 59, 65, 66, 68, 70, 71
    Head of, 26, 71
    Mouth of, 9, 27, 28, 41, 43, 66
Butterwood/Butterwood Spring, 1, 6, 36
Camp Creek, 10
Cellar/Celler Creek, 3, 13, 18, 49
Coldwater Run, head of, 6
Coles Branch/ James Coles' Spring Branch, 29, 53
Craddock's Branch, 68
Deep Creek, 1, 2, 4, 11, 13, 14, 17, 18, 21, 23, 26, 34, 35, 45, 47, 49, 59, 61, 62, 63
Fighting Creek, 20, 28
    Mouth of, 28
Flat/Flatt Creek, 1, 3, 4, 6, 7, 8, 10, 13, 14, 15, 16, 17, 18, 19, 20, 23, 24, 27, 28, 30, 31, 32, 33, 34, 35, 38, 39, 40, 41, 45, 47, 48, 49, 50, 51, 52, 53, 56, 62 64, 71, 72
    Head of, 71
    Mouth of, 28
Genito/Gennitoe/Jenito, 46, 48, 53, 57, 64
Great Nottoway River, 9, 15, 42, 50, 57, 59, 60
Great Saylor, 38
Hall Creek, 3
    Harrcane/Harricain/Harricane/Harrycain, 12, 15
Hurts Creek, 19, 28
Jenito [see Genito]
Judis Branch, 27
Knibbs/Knibs Creek [see also Nibbs/Nibs Creek], 4, 16, 17, 35, 39, 44, 62, 72
    South fork of, 44
Lazaretta/Lazaretto, 6, 25
    Head of, 25
Leith Creek, 23

Little Bryer River, 37
Little Creek, 42
Little Flatt Creek, head of, 19
Little Let Alone, 28
Little Nottoway, 2, 18, 36, 41, 42, 46, 50, 58, 60
Little Roanoak, 27, 54
    Head of, 27
Maherrin River, head of, 26
Malarys/Malereys/Mallarys/Mallerys/Mallorys Creek, 25, 26, 30, 35, 36, 47, 49, 61, 71
    Head of, 56
Mountain Creek, 11, 27, 56, 59, 70, 71
Namoseen/Namaszeen/Namazeen/Namozain/Namozn/Namozeen Creek, 1, 2, 3, 6, 7, 16, 28, 29, 34, 53, 62, 71, 72
Nibbs/Nibs Creek, 30, 63, 69
Nottoway River, 2, 6, 8, 12, 15, 16, 20, 21, 23, 25, 31, 35, 36, 42, 47, 48, 52, 59, 60, 71
    Fork of, 15
Old pond/Old ponds, 7, 8
Old ponds of Flatt Creek, 3
Peters Creek, 50
The (Appomattox?) River, 2, 12, 16, 19, 21, 22, 27
Rocky/Rockey Run, 18, 19, 29, 30, 71
Sailors Creek, 39, 40, 64, 65
Sandy Creek, 8, 10, 17, 19, 31, 50, 51, 55, 64
Sandy River, 9, 22, 30, 37, 48, 49, 53, 54, 59, 64, 65, 71
    Mouth of, 9
    Upper fork of, 49
Sawneys Creek, 53, 67
    Mouth of, 67
Saylors/Saylers Creek, 4, 8, 10, 19, 20, 21, 29, 36, 50, 51, 68
    Forks of, 4
Sellar/Sellers Creek, 46, 49
Smacks/Smax Creek, 2, 3, 10, 13, 42
Smiths Creek, 64
Snails/Snales Creek, 21, 33, 35, 37, 51, 58, 61
    First branch of, 21

    Head of, 51, 58
Spencers Branch, 29
Spiners/Spinners Run, 16, 62
Spring Creek, 31, 35
Stocks/Stokes Creek, 4, 8, 10, 17, 18, 27, 31, 32, 33, 49, 51, 59, 62
    Forks of, 4
Sweathouse/Swethouse/Swett House Creek, 1, 67, 70
Tallys Branch, 29
Tomahitton Creek, 4
Vaughans/Vaughns Creek, 17, 36, 65, 67
    Hills fork on, 17
West/Wests Creek, 1, 2, 3, 5, 7, 8, 13, 15, 20, 22, 27, 29, 33, 35, 36, 40, 42, 43, 47, 48, 50, 53, 55, 58, 59, 61, 64, 66, 67
Whetstone 50
Wintercomacke/Wintercomake/ Wintercomeck/Winticomak/Winticomake/ Wintocomack/Wintocomacke/Wintocomake/Wintycomeck/ Witocomeck creek, 2, 9, 10, 11, 16, 28, 29, 34, 40, 49, 53, 54 58, 62, 66, 67, 71, 72
Woody Creek, 20, 70

Roads

[NOTE: Roads are cross-indexed to all locations and persons mentioned. Descriptions have been standardized to aid in identifying roads and to simplify the preparation of this index. This was a necessity since many road descriptions changed slightly in the orders as different landmarks were cited. Various roads went under similar general descriptions and the reader should bear this in mind when determining the identify of each road.]

Road from Great Nottoway (Brunswick County) and along the road already cleared to the line between Amelia County and Prince George County (Road from Brunswick County to Prince George County), 9

Road from Andrews Branch to Nibbs Creek and thence from the fork below Anderson's to Booker's road, 63

Road from the Court House to the fork below Anderson, 64

Road from Anderson's bridge down, 52

Road from Flatt Creek and down Anderson's road, 1

Road across Beaver Pond Branch to Anderson's road, 1

Road from the Court House to Anderson's road near the race paths, 3, 7, 39

Road from the Court House into Anderson's road, 7, 39

Road from Anderson's road to Mr. Booker's road to the River bridge, 12

Road from William Neale's house to Anderson's road, 13

Road from Ward's quarter to Anderson's road, 14

Road from Anderson's road down to the bridge over the river, 16

Road from the Lawyers path to Anderson's road, 21

Road from Booker's fork to Anderson's road and to Bevill's bridge, 42

Road from Goode's bridge to Tanner's road thence to Anderson's road and to Thomas Jones's, 59

Road from the fork of Anderson's road down to Webster's commonly called Tanner's road, 67

Bridle way from the road near Charles Anderson's to Bush River Church, 43

Road from Charles Anderson's to the extent of the county towards Little Roanoak bridge, 54

Francis Anderson's road (continued to Stocks Creek), 49

Road near Harry Anderson's house, 3

Road from Flatt Creek Church to Mr. Henry Anderson's, 30

Road from Deep Creek to the road near James Anderson's, 23

Road beginning near James Anderson's and to Butterwood road at or near Leith Creek, 23

Road from the head of James Anderson's road to the head of Coldwater Run upon the ridge between Nottoway and the Lazaretta, 6

Road from James Anderson's road to Jordan's road to Nottoway Chapel, 9

Road from Leigh's bridge to the old road below James Anderson, Sr.'s house, 60

Road from the Court House to the cross roads below Mr. Anderson's and from thence to the Church, 47

Road from Mr. Anderson's bridge to Deep Creek bridge, 2

Road from Flatt Creek to Mrs. Anderson's quarter, 4

Road from Knibbs Creek to Mrs. Anderson's bridge to Mr. Booker's road, 16

Road from the bridge below Mrs. Anderson's quarter to Buckskin, 1

Road from Thomas Anderson's house to Thomas's road, 18

Road from Booker's fork to Andrews Branch, 63

Road from Andrews Branch to Nibbs Creek and thence from the fork below Anderson's to Booker's road, 63

Road from Knibs Creek to Andrews bridge, 35

NOTE: Entries for variant spellings of Appomatox are combined; all entries for Appomatox (and spelling variants) are assumed to indicate the Appomatox River

Road to Appomattox River near Jeneto (Jenneytoe), 8

Road from Thomas Tabb's house to the road over Appomatox River bridge, 13

Road from Knibbs Creek to the bridge over Appomatox River, 16

Road from Appomatox River near Col. Richard Randolph's quarter up to Hill's fork on Vaughn's Creek, 17

Road from Flatt Creek to Appomatox crossing West Creek, 20

Road from the fork of Booker's road to the lower bridge over Appomatox, 21

Road from the Ridge road to the place where the bridge over Appomatox River at Bass's will be built, 21

Road from William Tinstall's house to the road leading to Appomatox, 24

Road from the head of Little Ronoak along the ridge between Briery and Buffalloe Rivers to Rutlidge's ford over Appomattox River, 27

Road to Appomatox bridge, 28

Road from Sandy ford on Appamatox River to the main branch of Spring Creek, 31

Road from near Mr. Nash's house to Appomatox River a little above Bush River, 41

Road from the road near Mr. Nash's to the bridge over Appomatox River, 43

Road from Appomattox River to the County line, 50

Road from Appomattox River road that leads from Jeneto to Bush River, 57

Road from the ridge at the head of Appomattox River joining Callaway's road to where it crosses the river, 65

Road from the fork to Rutledge's ford in Appomattox River above the mouth of Bush River, 66

Road from Atwood's plantation on Bryery River to Roanoak road, 58

Road from James Atwood's road into Roanoak road, 53

Road from Abram Baker's mill path to the upper Botton through Mr. Wimbush's and thence to the County line, 58

Road from Baldwin's to the fork above Cheatham's, 62

Road from the fork above Baldwin's ordinary to James Cheatham's, 59

Road from Watson's race ground to Baldwin's/Baldwin's ordinary, 62, 71

Road from the Ridge road to the place where the bridge over Appomatox River at Bass's will be built, 21

Road from Peter Wynn's road to William Jackson's new road at Bates's path, 44

Road from Batts'/Battes' path to the County line, 39, 43

Road across Beaver Pond Branch to Anderson's road, 1

Road from Bush River road at Beisley's path crossing the creek below Watson's mill thence into Watson's road below Tunstall's quarter, 37

William Belcher's road (see also Tanner's road), 48

Road from the Bent Creek to Booker's road, 3

Road from Thomas Bevil's fork to Old Saponey ford, 31

[Fork of the road at] Bevill's at Bottom's, 72

Road from Booker's fork to Anderson's road and to Bevill's bridge, 42

Road from Deep Creek lower bridge to the fork of Booker's road above Boram's and thence to Bevill's bridge, 3

Road from Deep Creek bridge to the River bridge near Burton's and Bevill's lines, 21

Road from the County line between Tomahitton and the Birchen Swamps to the Chapel on Nottoway, (Road from Nottoway Chapel to Prince George County line) 4, 15, 59

Road from Maj. Bland's quarter to Dandy's race paths, 31

Road from Booker fork to the fork of the road leading to the upper River bridge, 22

Road from Booker's fork to Flatt Creek bridge, 34

Road from Booker's fork to Anderson's road and to Bevill's bridge, 42

Road from Booker's fork to Andrews Branch, 63

Road from Booker's mill into Lyles's road to the Court House, 7

Road from Booker's mill to the fork of the road at Good's bridge, 45

Road from the fork below [Cobbs'] ordinary to Booker's mill, 63

Road from Booker's mill to the fork at the school house, 63

Road from the fork of Booker's road to the Church, 2

Road from the Bent Creek to Booker's road, 3

Road from the fork of Booker's road to the lower bridge over Appomatox, 21

Road from Deep Creek lower bridge to the fork of Booker's road above Boram's and thence to Bevill's bridge, 63

Road from the fork of Booker's road to the fork of the road thence to Goode's bridge, 63

Road from Andrews Branch to Nibbs Creek and thence from the fork below Anderson's to Booker's road, 63

Road from the road to John Booker's crossing by his Brick-yard Branch to the main road to Col. Booker's, 34

Road from Good's bridge to Col. Booker's, 35

Road near Col. Booker's house, 39

Road between Thomas Spencer's and Edmund Booker, Jr.'s into the main road from the Court House to Warwick near the south fork of Knibbs Creek, 44

Road from Edward Booker's house to the Church, 7

Road from Richard Booker, Gent.'s mill to the new road to Mr. Edward Booker's, 16

Road near Edward Booker, Gent.'s house to his cleared ground, 22

Road through Edward Booker, Jr.'s land near Nottoway, 25

Road through Edward Booker Jr.'s land, 37

Road near George Booker's house, 28

Road near John Booker's house that is the way to Richard Booker's mill, 33

Road from the road to John Booker's crossing by his Brick-yard Branch to the main road to Col. Booker's, 34

Road to Mr. Booker's mill, 13

Road from Anderson's road to Mr. Booker's road to the River bridge, 12

Road from Knibbs Creek to Mrs. Anderson's bridge to Mr. Booker's road, 16

Road from Richard Booker's mill to Smax Creek, 13

Road from (Richard) Booker's mill to the fork of the road leading to the River bridge, 19, 27

Road for carts near Maj. Richard Booker's mill, 25

Road over Richard Booker's Millstream, 26

Road near John Booker's house that is the way to Richard Booker's mill, 33

Road from Richard Booker, Gent.'s mill to the new road to Mr. Edward Booker's, 16

Road from Deep Creek lower bridge to the fork of Booker's road above Boram's and thence to Bevill's bridge, 63

[Fork of the road at] Bevill's at Bottom's, 72

Forks [of the road] below Bottom's, 72

Road from Thomas Bottom's on West Creek to the Old Ponds of Flatt Creek along or near the old Ridge path, 3

Road from Abram Bakers mill path to the upper Botton through Mr. Wimbush's and thence to the County line, 58

NOTE: Entries for Boush River are combined with Bush River

Road from Bush River road a little below John Braggs' to the Church, 27

Road from William Watson's into the road at Brathwait's, 49

Road from the road to John Booker's crossing by his Brick-yard Branch to the main road to Col. Booker's, 34

NOTE: Entries for Briery River are combined with Bryery River

Road from William Brown's house to the Church, 13

Road from Browne's ordinary to Malary's Creek, 49

Road from Col. Richard Randolph's quarter to the ridge which divides this County from Brunswick, 6

Road from Great Nottoway (Brunswick County) and along the road already cleared to the line between Amelia County and Prince George County (Road from Brunswick County to Prince George County), 9

Road from Col. Richard Randolph's near Harden's to a road cleared by order Brunswick Court, 10

Road from John Hudson's to Brunswick County line, 11

Road to Brunswick County line, 14

NOTE: Entries for Briery River are combined with Bryery River

Road from the head of Little Ronoak along the ridge between Briery and Buffalloe Rivers to Rutlidge's ford over Appomattox River, 27

Road from Atwood's Plantation on Bryery River to Roanoak road, 58

Road from Watson's to the head of Bryery River, 59

Road from Bryery River to George Moore's, 64

Road from the bridge below Mrs. Anderson's quarter to Buckskin, 1

Road from West Creek to Buckskin, 20

Road from Dabney's to Buckweding, 5

Road from Buckweding, 33

NOTE: Entries for variant spellings of Buffalo are combined; entries for Buffalo (and spelling variants) are assumed to indicate the Buffalo River

Road from George Walker's plantation to Buffalo River, 5

Road from the head of Little Ronoak along the ridge between Briery and Buffalloe Rivers to Rutlidge's ford over Appomattox River, 27

Road from Hudson's ford on Buffilloe up the ridge opposite to Col. William Randolph's upper quarter, 34

Road crossing over Buffilloe, 43

Road from the race paths at Abraham Wawmock's to Sandy River bridge from thence to Bush River bridge and into the Buffelow road, 53

Road from Sailors Creek bridge to the fork of the road above Sandy River bridge to Bush River bridge and from thence to Buffelow road, 65

Road from Richard Stone's to the head of Buffelow-Bed Creek, 60

Road from Guilintine's new road to the road that comes down by Burton's, 61

Road from the Folly to the fork above Burton's, 64

Road from the Deep Creek bridge into the main road and to Burton's bridge, 17

Fork below Burton's bridge, 20

Road from Burton's bridge to the Church, 49

Road from Flatt Creek at Burton's bridge to Smith's Creek bridge at Winterham and from James Ferguson's to the Church, 64

Road from Deep Creek bridge to the River bridge near Burton's and Bevill's lines, 21

Mr. Walker's road from Saylors Creek to Crawford's and thence into Burton's road to the Court House, 8

Road from Crafford's house to Burton's road over Flatt Creek, 15

Road from the fork of Burton's road to Knibbs Creek, 16

Road from John Burton's house to the Court House, 64

Road by Mr. Burton's, 38

NOTE: Entries for Boush River are combined with Bush River

Road from Craddock's on Flatt Creek to Bush River, 4

Road from the mouth of Boush River below the mouth of Sandy River to Walker's road, 9

Road from Mr. Nash's quarter on Boush River to Osborn's road, 14

Boush River/Bush River road, 14, 35

Road from Boush River to Saylers Creek, 20

Road from Mallary's Creek along the ridge to Randolph's road at the head of Boush and Maherrin Rivers, 26

Road from near Mr. Nash's house to Appomatox River a little above Bush River, 41

Road from Griffin's road to Bush River, 47

Road from Appomattox River road that leads from Jeneto to Bush River, 57

Road from Sandy River to Nash's mill on Bush River, 59

Road from the fork to Rutledge's ford in Appomattox River above the mouth of Bush River, 66

Road from the head of Flatt Creek to head of Bush River, 71

Road from Boush River bridge across Saylors Creek to Walker's road, 21

Road from Bush River bridge to the Chapel, 29

Road from Bush River bridge into the road leading to Rutledge's ford, 34

Road from the race paths at Abraham Wawmock's to Sandy River bridge from thence to Bush River bridge and into the Buffelow road, 53

Road from Sailors Creek bridge to the fork of the road above Sandy River bridge to Bush River bridge and from thence to Buffelow road, 65

Road from Bush River bridge to the Church, 66

Bridle way from the road near Charles Anderson's to Bush River Church, 43

Road from Flatt Creek Down to Bush River fork, 56

Road from the fork road to Boush River road near Cheatham's, 19

Bridle way from Boush River road above Mr. Read's into Nottoway road, 21

Road from West Creek to Bush River road, 22

Bridle way/road from Bush River road to Irby's road to the Court House, 22, 25

Road from Bush River road a little below John Braggs' to the Church, 27

Road from Bush River road a little below the Pole bridge along the ridge into Mallory's Creek road, 36

Road from Bush River road at Beisley's path crossing the creek below Watson's mill thence into Watson's road below Tunstall's quarter, 37

Road from Sandy River where Capt. Walker's old road crossed it to Bush River road, 37

Bush River road above Cheatham's into the road near the south fork of Knibbs Creek, 44

Bridle way from the head of Snails Creek into Bush River road to Sandy River Chapel, 51

Road from Flatt Creek bridge at Mayes's to the fork of Bush River road near Walters's road, 53, 56

Road from Butterwood to the race paths near Mr. Irby's, 1

Road from Letbetter's low grounds on Nottoway River to Butterwood road, 6

Road beginning near James Anderson's and so to Butterwood road at or near Leith Creek, 23

Road from Nottoway bridge to Butterwood Spring into Cock's road, 36

Road from the ridge at the head of Appomattox River joining of Callaway's road to where it crosses the river, 65

Road from the Celler to Dandy's race paths, 3

Road from Chambers' to Neal's race paths, 36

Road from West Creek road just below Hudson's race paths down to Hugh Chambers' plantation, 36

Bridle way from West Creek to the Chapel on Nottoway, 2

Road from the County line between Tomahitton and the Birchen Swamps to the Chapel on Nottoway, (Road from Nottoway Chapel to Prince George County line) 4, 15, 59

Road from Harricain to the Chapel, 15

Road from Bush River bridge to the Chapel, 29

Road from Henry Robertson's mill path to Crenshaw's ford over Little Nottoway into Jordan's road below the Chapel, 36

Road from the bridge over Nottoway River to the Chapel road, 31

Bridle way from Little Bryer River into the Chapel road, 37

Road from the fork road to Boush River road near Cheatham's, 19

Bush River road above Cheatham's into the road near the south fork of Knibbs Creek, 44

Road from Cheatham's to the foot of the hill over Flatt Creek, 45

Road from Baldwin's to the fork above Cheatham's, 62

Road from the fork above Cheatham's to the Widow Mayes's, 71

Road from James Cheatham's to Judis Branch, 27

Old road by James Cheatham's, 46

Road from the fork above Baldwinn's ordinary to James Cheatham's, 59

Road from the old road above James Cheatham's to Knibbs Creek, 62

Road from the fork of Green's road to a Chestnut Oak, 62

Road from Namozain road to a Chestnut Oak on William Green's land, 34

Road from a Chestnut Oak to Deep Creek lower bridge, 62

Road from the Chestnut Oak to Deep Creek bridge, 34

Bridle way from the Rattlesnake ford to the Church on Flatt Creek, 1

Road from the fork of Booker's road to the Church, 2

Road from Flatt Creek to the Church, 6

Road from Edward Booker's house to the Church, 7

Road from the Church to Stocks Creek and continued to Sandy Creek, 8

Road from the Church to Stocks Creek, 8, 33

Road from the Church to West Creek, 13

Road from William Brown's house to the Church, 13

Road from Bush River road a little below John Braggs' to the Church, 27

Road from the new road near Mallary's Creek to Sandy River road to the Church, 30

Road from the Court House to the cross roads below Mr. Anderson's and from thence to the Church, 47

Road from Burton's bridge to the Church, 49

Road from the Church to Ferguss's bridge, 55

Road from Randolph's road below his lower quarter to the upper Church in Nottoway Parish, 56

Bridle way through the lands of John Farley, Isaac Morris and Stewart Farley into the main road to the Church, 57

Road from Flatt Creek at Burton's bridge to Smith's Creek bridge at Winterham and from James Ferguson's to the Church, 64

Road from Bush River bridge to the Church, 66

Road from the Church to the old road, 66

Road from Craddock's Branch to the Church, 68

Road from Mr. Cock's mill into the Church road, 9

Road from Great Nottoway to the Prince George County line and the Church road up to the Harrycain, 15

Crossroads (fork road?) where Jackson's road comes into the Church road at Nottoway, 20

The Church road, 30, 44

Bridle way from George Steegall's house to the Church road, 67

Road from the Church road to the Court House, 68

Road from William Clarke's into the main road, 7

Road from Robert Stoker's to Clayborn's, 54

Road to Clement's mill, 24

Road from Clement's old mill to Gillintine's, 32

Road from the causeway of Farguson's bridge to Mr. Clement's, 61

William Clement's road, 6

Road from Flatt Creek lower bridge to Col. Cobbs' ordinary, 35

Road from the fork below [Cobbs'] ordinary to Booker's mill, 63

Road from Cobbs'/Col. Cobbs' ordinary to the Court House, 63, 72

Road from Nibbs Creek to Col. Cobbs' ordinary, 69

The first fork above the fork of the road to Warwick at Col. Cobbs' ordinary, 72

Road to Col. Cobbs' ordinary, 72

Cross roads at Col. Cobbs' ordinary, 72

Road from Flatt Creek bridge on Lisles's road to Mr. Cobbs' ordinary, 63

Road by Samuel Cobbs' house, 42

Fork of Cock's and Irby's roads, 20

Road from Cock's fork to Farley's, 72

Road from Cock's quarter to West Creek, 27

Road from Nottoway bridge to Butterwood Spring into Cock's road, 36

Cock's road, 57

Road from Mr. Cock's mill into the Church road, 9

Road from Jordan's bridge to Cock's/Mr. Cock's road, 9, 15, 60

Road from the old road into Mr. Cock's road, 43

Road from Mr. Cock's road to Nottoway road, 60

Road from the head of James Anderson's road to the head of Coldwater Run upon the ridge between Nottoway and the Lazaretta, 6

Road from Namozain Creek/bridge to Coles's/James Coles's Spring Branch, 29, 53

Road from James Coles's Spring Branch to Wintercomake, 29

Road from Combs's bridge over Flatt Creek to the Court House, 16

Road to Combs's bridge, 12

Old road near John Comb's, 33

Road from the County line between Tomahitton and the Birchen Swamps to the Chapel on Nottoway, (Road from Nottoway Chapel to Prince George County line) 4, 15, 59

Road from Snales Creek to the County line, 35

Road from Batts'/Battes' path to the County line, 39, 43

Road from the County line to Jackson's/James Jackson's road, 40, 43

Road from Appomattox River to the County line, 50

Road from Abram Bakers mill path to the upper Botton through plantation Mr. Wimbush's and thence to the County line, 58

Road from the Harricane bridge to the County line, 60

Road from Charles Anderson's to the extent of the County towards Little Roanoak bridge, 54

Bridle way/road from West Creek to the Court House/Court House road, 2, 7, 15, 61, 64

Road from the Court House to Anderson's road near the race paths, 3, 7, 39

Road from Flatt Creek to the Court House, 3, 7, 17, 32

Road from Booker's mill into Lyles's road to the Court House, 7

Road from the Court House into Anderson's road, 7, 39

Mr. Walker's road from Saylors Creek to Crawford's and thence into Burton's road to the Court House, 8

Road from Stocks Creek to the Court House, 10

Bridle way from Robert Vaughan's to the Court House, 11

Road from Combs's bridge over Flatt Creek to the Court House, 16

Bridle way/road from Bush River road to Irby's road to the Court House, 22, 25

Road from Edward Jones's to cross West's Creek near Tally's Branch then between Wilkinson's quarter and Ward's to the Court House, 29

Road from William Yarbrough's house to Capt. Irby's road to the Court House, 29

Road between Thomas Spencer's and Edmund Booker, Jr.'s into the main road from the Court House to Warwick near the south fork of Knibbs Creek, 44

Road from Southall's to the Court House, 47

Road from the Court House to the cross roads below Mr. Anderson's and from thence to the Church, 47

Road from Cobbs'/Col. Cobbs' ordinary to the Court House, 63, 72

Road from John Burton's house to the Court House, 64

Road from the Court House to the fork below Anderson, 64

Road from Jeneto road to the Court House, 65

Road from the Church road to the Court House, 68

Road from the Court House to George Elliotts, 70

Road from Nottoway Church along Court House road to West Creek, 59

Road from the Court House road to Griffin's road, 60

Road from the race paths to the Court House road, 61

Road from the Court House road to West Creek road below Hudson's race paths, 67

Road from Tanner's to Craddock's, 4

Road from Craddock's on Flatt Creek to Bush River, 4

Road from Jeneto road to a branch near Craddock's plantation, 68

Road from Craddock's Branch to the Church, 68

Road from Ward's quarter to the foot of the hills the other side Flatt Creek and Craddock's bridge, 14

Road from Craddock's bridge upward, 46

Road from the Sand road into Craddock's road, 8

Road to Crafford's from John Hudson's, 11

Saylors Creek road turned at Charles Johnson's into the road from Crafford's to Dawson's race paths   51

Road from Crafford's house to Burton's road over Flatt Creek   15

Mr. Walker's road from Saylors Creek to Crawford's and thence into Burton's road to the Court House, 8

Road from Great Saylor into the road a little below Crawford's house   38

Road from Flatt Creek/Flatt Creek bridge to Crawford's house, 56, 62

Road from Crawford's house to Great Sailors Creek bridge, 65

Road from Crawford's house to the Folly, 66

Road from Saylors Creek road to the road that goes to Crawford's house, 68

Road from Henry Robertson's mill path to Crenshaw's ford over Little Nottoway into Jordan's road below the Chapel, 36

Crossroads (fork?) where Watson's road comes to Flatt Creek, 20

Road from the Court House to the cross roads below Mr. Anderson's and from thence to the Church, 47

Road from the cross road to West Creek, 61

Cross road near George Hamm's, 72

Cross roads at Col. Cobbs' ordinary, 72

[Cross road] near Good's bridge, 72

Road from Dabney's to the Cutt banks, 2

Road from the Cutt Banks, 7

Road from Dabney's to the Cutt banks, 2

Road from Dabney's to Buckweding, 5

Road from the road by Dandy's into the road near Phillip Pledger's, 56, 60

Road from the Celler to Dandy's race paths, 3

Road from Dandy's race paths to Capt. Stark's new quarter, 16

Road from Spinner's to Dandy's race paths, 16

Road from the Rev. Mr. John Ornsby's to Dandy's race paths, 26

Road from Dandy's race paths to Thomas Jones's road, 26

Road from Maj. Bland's quarter to Dandy's race paths, 31

Peter Davis' road, 39

Saylors Creek road turned at Charles Johnson's into the road from Crafford's to Dawson's race paths, 51

Road from Deep Creek, 4

Road from Deep Creek to Knibbs Creek, 4

Road from Deep Creek to the road near James Anderson's, 23

Road from where Irby's road crosses the road to Mayes's down to Deep Creek 26

Road from West Creek to Deep Creek, 61

Road from Smax Creek to the river and Deep Creek bridge, 2

Road from Mr. Anderson's bridge to Deep Creek bridge, 2

Road from the Deep Creek bridge into the main road and to Burton's bridge, 17

Road from Deep Creek bridge to the River bridge near Burton's and Bevill's lines, 21

Road from the Chestnut Oak to Deep Creek bridge, 34

Road from Deep Creek bridge to Capt. Irby's cross road, 35

Road from Flatt Creek to Sappony ford and Deep Creek lower bridge, 1

Road from Deep Creek lower bridge to Namozeen road, 1

Road from a Chestnut Oak to Deep Creek lower bridge, 62

Road from Deep Creek lower bridge to the fork of Booker's road above Boram's and thence to Bevill's bridge, 63

Deep Creek road, 11

Road near Daniel Degarnett's, 42

Road from the head of Yarborrough's road up the ridge between Peters Creek and Whetstone to Tukaer's cart path then along the path up the ridge between Little Nottoway and Great Nottoway to the road near Degernett's, 50

Road from Mallory's Creek to Degernett's, 71

Road along the ridge from Moore's below Dejarnett's Smith Shop near head of Snails Creek, 58

Road from Richard Dennis' above Woody Creek, 70

Road from William Eckhols's road on Stocks Creek up to the ridge at the fork of Sandy Creek, 17

Road from the Court House to George Elliotts, 70

Road from the causeway of Ferguson's bridge to Mr. Clement's, 61

Road from Cock's fork to Farley's, 72

Fork of the roads near Henry Farley's, 72

Bridle way through the lands of John Farley, Isaac Morris and Stewart Farley into the main road to the Church, 57

Road on both sides of Ferguson's bridge, 49

Road from Flatt Creek at Burton's bridge to Smith's Creek bridge at Winterham and from James Ferguson's to the Church, 64

Road from the Church to Ferguss's bridge, 55

Road beginning a little below John Winn's to Fisher's cart path and from thence to Jordan's bridge, 14

The five forks, 72

Road from Flatt Creek to Sappony ford and Deep Creek lower bridge, 1

Road from Flatt Creek and down Anderson's road, 1

Bridle way from the Rattlesnake ford to the Church on Flatt Creek, 1

Road from Flatt Creek to the Court House, 3, 7, 17, 32

Road from Thomas Bottom's on West Creek to the Old Ponds of Flatt Creek along or near the old Ridge path, 3

Road from the White Oak on Flatt Creek to John Hurt's near the fork of Stocks Creek, 4

Road from Flatt Creek to the fork of Saylors Creek, 4

Road from Flatt Creek to Mrs Anderson's quarter, 4

Road from Craddock's on Flatt Creek to Bush River, 4

Road from Flatt Creek to the Church, 6

Road from Liles/Lyle's/Lyle's ford to Flatt Creek/Flatt Creek bridge, 6, 23, 52, 62

Road from Smacks Creek to Flatt Creek, 10

Road from Ward's quarter to the foot of the hills the other side Flatt Creek and Craddock's bridge, 14

Road from Crafford's house to Burton's road over Flatt Creek, 15

Road from Combs's bridge over Flatt Creek to the Court House, 16

Road beginning above and near the mouth of Stocks Creek into the main road over Flatt Creek, 18

Crossroads (fork?) where Watson's road comes to Flatt Creek, 20

Road from Flatt Creek to Appomatox crossing West Creek, 20

Road from Stocks Creek to Flatt Creek, 27

Road from Hurt's Creek to Flatt Creek, 28

Road from Flatt Creek/Flatt Creek bridge to Southall's/Southall's ordinary, 35, 47

Road from John Robertson's across Flatt Creek through the land of John Gibbs and Essex Worsham to the road to Good's bridge, 38

Road from Cheatham's to the foot of the hill over Flatt Creek, 45

Road from Flatt Creek/Flatt Creek bridge to Crawford's house, 56, 62

Road from Flatt Creek at Burton's bridge to Smith's Creek bridge at Winterham and from James Ferguson's to the Church, 64

Road from the head of Flatt Creek to head of Bush River, 71

Road from the head of Flatt Creek to George Moor's, 72

Road from Booker's fork to Flatt Creek bridge, 34

Road from Flatt Creek lower bridge to Col. Cobbs' ordinary, 35

Road across Flatt Creek upper bridge, 35

Road from Flatt Creek bridge at Mayes's to the fork of Bush River road near Walters's road, 53, 56

Road from Flatt Creek bridge to the fork below Foster's   62

Road from Flatt Creek bridge on Lisles's road to Mr. Cobbs' ordinary   63

Bridle way from the Rattlesnake ford to the Church on Flatt Creek, 1

Road from Flatt Creek Church to Mr. Henry Anderson's, 30

Road from Gillenten's to Flatt Creek Church, 47

Road from Sandy Creek to the Folly, 64

Road from the Folly to the fork above Burton's, 64

Road from Crawford's house to the Folly, 66

Road from Flatt Creek bridge to the fork below Foster's, 62

Franklin's road, 5

Geneto road (see Jeneto)

Road from John Robertson's across Flatt Creek through the land of John Gibbs and Essex Worsham to the road to Good's bridge, 38

NOTE: See entries for Guilintine in addition to Gillenten and Gillintine

Road from Gillenten's to Flatt Creek Church, 47

Road from Gillintine's/Gullington's to Stocks Creek, 32

Road from Clement's old mill to Gillintine's, 32

Road from Good's bridge to Col. Booker's, 35

Road from John Robertson's across Flatt Creek through the land of John Gibbs and Essex Worsham to the road to Good's bridge, 38

Road from John Robertson's ford into the road to Good's bridge below Peter Webster's, 45

Road from Booker's mill to the fork of the road at Good's bridge, 45

Road from Goode's bridge to Warwick, 48

Road from Goode's bridge to Tanner's road thence to Anderson's road and to Thomas Jones's, 59

Road from the fork of Booker's road to the fork of the road thence to Goode's bridge, 63

Road from Murray's path to Goode's bridge, 63

[Cross road] near Good's bridge, 72

Road from Granger's path to the road to the bridge, 12

Road from Great Nottoway (Brunswick County) and along the road already cleared to the line between Amelia County and Prince George County (Road from Brunswick County to Prince George County), 9

Road from Jordons bridge to Great Nottoway, 15

Road from Great Nottoway to the Prince George County line and the Church road up to the Harrycain, 15

Road from the head of Yarborrough's road up the ridge between Peters Creek and Whetstone to Tukaer's cart path then along the path up the ridge between Little Nottoway and Great Nottoway to the road near Degernett's, 50

Road from Great Nottoway at Hampton Wade's to Nottoway bridge, 59

Road from Great Nottoway/Great Nottoway bridge to Little Nottoway/Little Nottoway bridge, 42, 60

Road from Crawford's house to Great Sailors Creek bridge, 65

Road from Great Saylor into the road a little below Crawford's house, 38

Road from Green's road to Wintocomeck bridge, 62

Road from Namazeen bridge to the fork of Green's road, 62

Road from the fork of Green's road to a Chestnut Oak, 62

Bridle way from Abraham Green's house to Rockey Run Chapel, 30

Road from Namozain road to a Chestnut Oak on William Green's land, 34

Road from Griffin's road to Bush river, 47

Road from the Court House road to Griffin's road, 60

Bridle way from Anthony Griffin's road to Sandy River Chapel, 22

NOTE: See entries for Gillenten and Gillintine in addition to Guilintine

Road from Guilintine's new road to the road that comes down by Burton's, 61

Road from Edmund Walker's to the road by Nicholas Guilintine's, 61

Road from Hall Creek to Smax Creek, 3

Road from George Ham's into the road to Saylors Creek, 36

Cross road near George Hamm's, 72

Road from Col. Richard Randolph's near Harden's to a road cleared by order Brunswick Court, 10

NOTE: Entries for variant spellings of Harricane (Hurricane Creek) are combined

Road from Harrcane to Jordan's road, 12

Road from Harricain to the Chapel, 15

Road from Great Nottoway to the Prince George County line and the Church road up to the Harrycain, 15

Road from Nottoway bridge to the Harricane bridge, 60

Road from the Harricane bridge to the County line, 60

Road from George Moore's house to the Hatters, 52

Road from Appomatox River near Col. Richard Randolph's quarter up to Hill's fork on Vaughn's Creek, 17

Road from Christopher Hinton's to the main road below Rocky Run Chapel, 19

Road from Hudson's on West Creek down to Mann's road, 33

Hudson's cart way, 35

Road from Hudson's ford on Buffilloe up the ridge opposite to Col. William Randolph's upper quarter, 34

Road from Hudson's ford to the road to Rutledge's ford near the school house, 34

Road from West Creek road just below Hudson's race paths down to Hugh Chambers' plantation, 36

Road from the Court House road to West Creek road below Hudson's race paths, 67

Hudson's road, 40

Road from John Hudson's to Brunswick County line, 11

Road to Crafford's from John Hudson's, 11

Road from Hurt's Creek to Flatt Creek, 28

Road from the White Oak on Flatt Creek to John Hurt's near the fork of Stocks Creek, 4

Fork of Cock's and Irby's roads, 20

Road from Yarbrow's to Woody Creek crossing Irby's road, 20

Bridle way/road from Bush River road to Irby's road to the Court House, 22, 25

Road from West Creek to Irby's road, 47

Road from where Irby's road crosses the road to Mayes's down to Deep Creek, 26

The Parsons road from Watson's road into Irby's road, 47

Road from Deep Creek bridge to Capt. Irby's cross road, 35

Road from William Yarbrough's house to Capt. Irby's road to the Court House, 29

Road from Charles Irby's house to West Creek, 15

Road from Butterwood to the race paths near Mr. Irby's, 1

Crossroads (fork road?) where Jackson's road comes into the Church road at Nottoway, 20

Road from the County line to Jackson's/James Jackson's road, 40, 43

Road from Stoker's bridge to Jackson's road, 60

Road from John Jackson's to the fork of Little Nottoway, 2

Road from Peter Wynn's road to William Jackson's new road at Bates's path, 44

Road from Stony bridge to Jeneto bridge, 64

Road to Appomattox River near Jeneto (Jenneytoe), 8

Road from Jeneto road to the new bridge, 50

Road from Sandy Creek to the fork of the road to Jeneto, 55

Road from Appomattox River road that leads from Jeneto to Bush River, 57

Road from Jeneto road to the Court House, 65

Road from Jeneto road to a branch near Craddock's plantation, 68

Saylors Creek road turned at Charles Johnson's into the road from Crafford's to Dawson's race paths, 51

Road over the Swethouse Creek below Abraham Jones's quarter to the main road, 1

Bridle path through Maj. Munford's land to Abraham Jones's mill, 46

Road from Capt. Jones's quarter to Wintercomake, 11

Road round Daniel Jones's fence, 58

Road from Watson's race ground to the fork below Daniel Jones's, 59

Road from Edward Jones's to cross West's Creek near Tally's Branch then between Wilkinson's quarter and Ward's to the Court House, 29

Road from West Creek to the fork of the road near Capt. Peter Jones's quarter/Maj. Peter Jones's, 1, 43

Road from Capt. Peter Jones's fork to Wintercomake, 2

[Cross road] near Maj. Peter Jones's, 72

Road over the head of Maj. Richard Jones's mill, 23

Road from Dandy's race paths to Thomas Jones's road, 26

Road from Goode's bridge to Tanner's road thence to Anderson's road and to Thomas Jones's, 59

Fork [of the road] where Thomas Jones formerly lived, 72

Road from Dandy's race paths to Thomas Jones's road, 26

Road beginning a little below John Winn's to Fisher's cart path to Jordan's bridge, 14

Road from Jordon's bridge to Cock's/Mr. Cock's road, 9, 15, 60

Road from Jordon's bridge to Great Nottoway, 15

Road from James Anderson's road to Jordan's road to Nottoway Chapel, 9

Road from Harrcane to Jordan's road, 12

Road from Henry Robertson's mill path to Crenshaw's ford over Little Nottoway into Jordan's road below the Chapel, 36

Road from Little Nottoway bridge in Jordan's road to William Maynard's, 58

Road from James Cheatham's to Judis Branch, 27

NOTE: See entries for Nibs/Nibbs Creek in addition to Knibs/Knibbs Creek

Road from Deep Creek to Knibbs Creek, 4

Road from Knibbs Creek to the bridge over Appomatox River, 16

Road from the fork of Burton's road to Knibbs Creek, 16

Road from Knibbs Creek to Mrs. Anderson's bridge to Mr. Booker's road, 16

Road from Knibs Creek to Andrews bridge, 35

Road between Thomas Spencer's and Edmund Booker, Jr.'s into the main road from the Court House to Warwick near the south fork of Knibbs Creek, 44

Bush River road above Cheatham's into the road near the south fork of Knibbs Creek, 44

Road from the old road above James Cheatham's to Knibbs Creek, 62

Road from the Lawyers path to Anderson's road, 21

Road from the head of James Anderson's road to the head of Coldwater Run upon the ridge between Nottoway and the Lazaretta, 6

Bridle way from the new road at the head of Lazaretto to Nottoway Chapel, 25

Road from Leigh's bridge to the old road below James Anderson, Sr.'s house, 60

Road beginning near James Anderson's and so to Butterwood road at or near Leith Creek, 23

NOTE: Entries for Liles/Lisles are combined with Lyles

Road from Letbetter's low grounds on Nottoway River to Butterwood road, 6

Bridle way from Little Bryer River into the Chapel road, 37

Road from the head of Little Flatt Creek to the fork road by Mr. Sherwin's plantation, 19

Road from John Jackson's to the fork of Little Nottoway, 2

Road from Henry Robertson's mill path to Crenshaw's ford over Little Nottoway into Jordan's road below the Chapel, 36

Yarbrough's road crossing Little Nottoway, 41

Road from the head of Yarborrough's road up the ridge between Peters Creek and Whetstone to Tukaer's cart path then along the path up the ridge between Little Nottoway and Great Nottoway to the road near Degernett's, 50

Road from Little Nottoway to where the new bridge is to be built, 58

Road from Great Nottoway/Great Nottoway bridge to Little Nottoway/Little Nottoway bridge, 42, 60

Road from Little Nottoway bridge in Jordan's road to William Maynard's, 58

Road from Thomas's road to Little Nottoway bridge, 60

Road from the head of Little Ronoak along the ridge between Briery and Buffalloe Rivers to Rutlidge's ford over Appomattox River, 27

Road from Charles Anderson's to the extent of the County towards Little Roanoak bridge, 54

Road from Snails Creek to Lunenburg line, 33

Road from Spring Creek to Lunenburg line, 35

NOTE: Entries for Liles/Lisles are combined with Lyles

Road from Flatt Creek/Flatt Creek Bridge to Liles/Lyle's/Lyle's ford, 6, 23, 52, 62

Road from Booker's mill into Lyles's road to the Court House, 7

Road from Lyles's ford, 20

Lyle's road, 51

Road from Flatt Creek bridge on Lisles's road to Mr. Cobbs' ordinary 63

Road from Mallary's Creek along the ridge to Randolph's road at the head of Boush and Maherrin Rivers, 26

Road from Thomas's to Main Nottoway, 20

Road round Samuel Major's Plantation, 47

Road from Stock/Stocks/Stokes Creek to Samuel Major's, 49, 59, 62

NOTE: All variant spellings of Mallory (i.e., Mallary, Mallery, etc.) are combined

Road from Mallery's Creek to the race paths at Watson's, 25, 61

Road from Mallary's Creek along the ridge to Randolph's road at the head of Boush and Maherrin Rivers, 26

Road from the new road near Mallary's Creek to Sandy River road to the Church, 30

Road from Snails/Snales Creek across Mallary's Creek, 35, 61

Road from Capt. Watson's to Malerey's Creek, 47

Road from Browne's ordinary to Malary's Creek, 49

Road from Capt. Watson's cart path to Mallery's Creek, 49

Road round the head of Mallory's Creek, 56

Road from Mallory's Creek to Degernett's, 71

Road from Bush River road a little below the Pole bridge along the ridge into Mallory's Creek road, 36

Road from Hudson's on West Creek down to Mann's road, 33

Road from John Martin's in Mr. Walker's road, 9

Road from where Irby's road crosses the road to Mayes's down to Deep Creek, 26

Road from Flatt Creek bridge at Mayes's to the fork of Bush River road near Walters's road, 53, 56

Road from the fork above Cheatham's to the Widow Mayes's, 71

Road from Little Nottoway bridge in Jordan's road to William Maynard's, 58

Road from the Old Ponds to William Mole's, 7

Road along the ridge from Moore's below Dejarnett's Smith Shop near head of Snails Creek, 58

Road near/by George Moore's house, 22, 37, 38

Road from Snails Creek to George Moore's house and to Col. Richard Randolph's upper quarter, 37

Road from George Moore's to Snails Creek, 37, 61

Road from George Moore's house to the Hatters, 52

Roads from Capt. Watson's to the fork above George Moore's, 54

Road from George Moore's to Sandy River/upper fork of Sandy River, 49, 54, 59, 64

Road from George Moore's to the fork of Randolph's road, 64

Road from Bryery River to George Moore's, 64

Road from the head of Flatt Creek to George Moor's, 72

George Moore's road, 39

Bridle way through the lands of John Farley, Isaac Morris and Stewart Farley into the main road to the Church, 57

Road near William Mott's house, 5

Road crossing Mountain Creek, 56

Bridle path through Maj. Munford's land to Abraham Jones's mill, 46

Road from Nibs Creek bridge to Murray's path, 63

Road from Murray's path to Goode's bridge, 63

NOTE: All variant spellings of Namozine (i.e., Namozain, etc.) are combined

Road from Namozeen bridge to the Swett house Creek, 1

Road from Wintocomake to Namaszeen/Namozain bridge, 16, 71

Road from Namozain Creek/bridge to Coles's/James Coles's Spring Branch, 29, 53

Road from Namazeen bridge to the fork of Green's road, 62

Road to Namozain Church, 31

Road from Deep Creek lower bridge to Namozeen road, 1

Road from Namozain road to a Chestnut Oak on William Green's land, 34

Road from Sandy River to Nash's mill on Bush River, 59

Road from near Mr. Nash's house to Appomatox River a little above Bush River, 41

Road from the road near Mr. Nash's to the bridge over Appomatox River, 43

Road from Saylors Creek to Mr. Nash's road above Womawk's, 51

Road from Mr. Nash's quarter on Boush River to Osborn's road, 14

Road from Neal's to Sappony road, 36

Road from Chambers' to Neal's race paths, 36

Road from William Neale's house to Anderson's road, 13

NOTE: See entries for Knibs/Knibbs Creek in addition to Nibs/Nibbs Creek

Road from Nibs Creek bridge to Murray's path, 63

Road from Andrews Branch to Nibbs Creek and thence from the fork below Anderson's to Booker's road, 63

Road from Nibbs Creek to Col. Cobbs' ordinary, 69

NOTE: In addition to entries for Nottoway, see also entries for Great Nottoway, Little Nottoway and Main Nottoway

Road from the head of James Anderson's road to the head of Coldwater Run upon the ridge between Nottoway and the Lazaretta, 6

Road from Letbetter's low grounds on Nottoway River to Butterwood road, 6

Road to Nottoway River, 8

Road from Nottoway road to the fork of Nottoway, 15

Road up to the ridge of Nottoway, 16

Road from Nottoway to West Creek (Nottoway road to West Creek), 20, 42, 47, 48, 58

Crossroads (fork road?) where Jackson's road comes into the Church road at Nottoway, 20

Road through Edward Booker, Jr.'s land near Nottoway, 25

Road from John Thomas's to the bridge from Nottoway and to ridge path near the race path, 20

Road from the bridge over Nottoway River to the Chapel road, 31

(Proposed) bridge where Winingham's road crosses Nottoway River, 35

Road from Nottoway bridge to Butterwood Spring into Cock's road, 36

Road from Great Nottoway at Hampton Wade's to Nottoway bridge, 59

Road from Nottoway bridge to the Harricane bridge, 60

Bridle way from West Creek to the Chapel on Nottoway, 2

Road from the County line between Tomahitton and the Birchen Swamps to the Chapel on Nottoway (Road from Nottoway Chapel to Prince George County line), 4, 15, 59

Road from James Anderson's road to Jordan's road to Nottoway Chapel, 9

Bridle way from the new road at the head of Lazaretto to Nottoway Chapel, 25

Road from Nottoway Church to West Creek, 55

Road from Watson's road to Nottoway Church, 59

Road from Nottoway Church along Court House road to West Creek, 59

Road from Randolph's road below his lower quarter to the upper Church in Nottoway Parish, 56

The Nottoway road, 4

Road from Nottoway road to the fork of Nottoway, 15

Bridle road from Nottoway road to Rocky Run Chapel, 18

Bridle way from Boush River road above Mr. Read's into Nottoway road, 21

Road from Mr. Cock's road to Nottoway road, 60

Road from the old pond to West Creek, 8

Road from Thomas Bottom's on West Creek to the Old Ponds of Flatt Creek along or near the old Ridge path, 3

Road from the Old Ponds to William Mole's, 7

Road from the Rev. Mr. John Ornsby's to Dandy's race paths, 26

Road from Mr. Nash's quarter on Boush River to Osborn's road, 14

The Parsons road from Watson's road into Irby's road, 47

Road from the head of Yarborrough's road up the ridge between Peters Creek and Whetstone to Tukaer's cart path then along the path up the ridge between Little Nottoway and Great Nottoway to the road near Degernett's, 50

Road from the road by Dandy's into the road near Phillip Pledger's, 56, 60

Road from Bush River road a little below the Pole bridge along the ridge into Mallory's Creek road, 36

Road from the head of Pole bridge down to the fork, 61

Road through John Pride's plantation, 50

Road from Great Nottoway (Brunswick County) and along the road already cleared to the line between Amelia County and Prince George County (Road from Brunswick County to Prince George County), 9

Road from the County line between Tomahitton and the Birchen Swamps to the Chapel on Nottoway, (Road from Nottoway Chapel to Prince George County line) 4, 15, 59

Road from Great Nottoway to the Prince George County line and the Church road up to the Harrycain, 15

Road from the Court House to Anderson's road near the race paths, 3, 7, 39

Road from John Thomas's to the bridge from Nottoway and to ridge path near the race path, 20

Road from or near Whitworth's in Saylors Creek road below the race paths, 29

Road from the race paths to the Court House road, 61

Road from the Celler to Dandy's race paths, 3

Road from Dandy's race paths to Capt. Stark's new quarter, 16

Road from Spinner's to Dandy's race paths, 16

Road from the Rev. Mr. John Ornsby's to Dandy's race paths, 26

Road from Dandy's race paths to Thomas Jones's road, 26

Road from Maj. Bland's quarter to Dandy's race paths, 31

Saylors Creek road turned at Charles Johnson's into the road from Crafford's to Dawson's race paths, 51

Road from West Creek road just below Hudson's race paths down to Hugh Chambers' plantation, 36

Road from the Court House road to West Creek road below Hudson's race paths, 67

Road from Butterwood to the race paths near Mr. Irby's   1

Road from Chambers' to Neal's race paths, 36

Road from Mallery's Creek to the race paths at Watson's, 25, 61

Road from Watson's race ground to the fork below Daniel Jones's, 59

Road from Watson's race ground to Baldwin's/Baldwin's ordinary, 62, 71

Road from the race paths at Abraham Wawmock's to Sandy River bridge from thence to Bush River bridge and into the Buffelow road, 53

Road from Randolph's road below his lower quarter to the upper Church in Nottoway Parish, 56

Road from Mallary's Creek along the ridge to Randolph's road at the head of Boush and Maherrin Rivers, 26

Road from Sawneys Creek into Randolph's road, 53

Road from George Moore's to the fork of Randolph's road, 64

Road from Vaughan's Creek to Randolph's road, 65

Old Rolling road from Randolph's road near his mill and thence across Vaughan's Creek at the old ford and then to continue the old way to the mouth of Sawneys Creek, 67

Road from Mr. Walker's plantation into the road from Col. Randolph's quarters, 5

Road from Sandy ford to Col. Randolph's mill and Vaughan's Creek, 36

Road from Col. Richard Randolph's quarter to the ridge which divides this County from Brunswick, 6

Road from Col. Richard Randolph's near Harden's to a road cleared by order Brunswick Court, 10

Road from Appomatox River near Col. Richard Randolph's quarter up to Hill's fork on Vaughn's Creek, 17

Road from Snails Creek to George Moore's house and to Col. Richard Randolph's upper quarter, 37

Road from Hudson's ford on Buffilloe up the ridge opposite to Col. William Randolph's upper quarter, 34

Bridle way from the Rattlesnake ford to the Church on Flatt Creek, 1

Road near/around William Ray's plantation, 51

Bridle way from Boush River road above Mr. Read's into Nottoway road, 21

Road from Col. Richard Randolph's quarter to the Ridge which divides this County from Brunswick, 6

Road from the head of James Anderson's road to the head of Coldwater Run upon the Ridge between Nottoway and the Lazaretta, 6

Road from William Eckhols's road on Stocks Creek up to the Ridge at the fork of Sandy Creek, 17

Road from Watson's muster field along the Ridge to the first branch of Snales Creek, 21

Road from Mallary's Creek along the Ridge to Randolph's road at the head of Boush and Maherrin Rivers, 26

Road from the head of Little Ronoak along the Ridge between Briery and Buffalloe Rivers to Rutlidge's ford over Appomattox River, 27

Road from Hudson's ford on Buffilloe up the Ridge Opposite to Col. William Randolph's upper quarter, 34

Road from Bush River road a little below the Pole bridge along the Ridge into Mallory's Creek road, 36

Road from the head of Yarborrough's road up the Ridge between Peters Creek and Whetstone to Tukaer's cart path then along the path up the ridge between Little Nottoway and Great Nottoway to the road near Degernett's, 50

Road along the Ridge from Moore's below Dejarnett's Smith Shop near head of Snails Creek, 58

Road from the Ridge at the head of Appomattox River joining Callaway's road to where it crosses the river, 65

Road from Thomas Bottom's on West Creek to the Old Ponds of Flatt Creek along or near the old Ridge path, 3

Road from John Thomas's to the bridge from Nottoway and to Ridge path near the race path, 20

Road from the Ridge road to the place where the bridge over Appomatox River at Bass's will be built, 21

The Ridge road, 27, 40, 41

Road from Smax Creek to the river and Deep Creek bridge, 2

Road from Anderson's road to Mr. Booker's road to the River bridge, 12

Road from Anderson's road down to the bridge over the River, 16

Road from (Richard) Booker's mill to the fork of the road leading to the River bridge, 19, 27

Road from Deep Creek bridge to the River bridge near Burton's and Bevill's lines, 21

Road from Booker fork to the fork of the road leading to the upper River bridge, 22

Road from James Atwood's road the nearest and best way into Roanoak road, 53

Road from Atwood's Plantation on Bryery River to Roanoak road, 58

Cart path from Henry Robertson/Robinson's house to his mill, 25, 26, 36

Road from Henry Robertson's mill path to Crenshaw's ford over Little Nottoway into Jordan's road below the Chapel, 36

Road from John Robertson's across Flatt Creek through the land of John Gibbs and Essex Worsham to the road to Good's bridge, 38

Road from John Robertson's ford into the road to Good's bridge below Peter Webster's, 45

Road from Rockey Run to Spencer's Branch, 29

Road from Wintercomacke to Rockey Run, 29, 71

Bridle road from Nottoway road to Rocky Run Chapel, 18

Road from Christopher Hinton's to the main road below Rocky Run Chapel, 19

Bridle way from Abraham Green's house to Rockey Run Chapel, 30

NOTE: Entries for Rutledge and Rutlidge are combined

Road from the head of Little Ronoak along the ridge between Briery and Buffalloe Rivers to Rutlidge's ford over Appomattox River, 27

Road from Hudson's ford to the road to Rutledge's ford near the school house, 34

Road from Bush River bridge into the road leading to Rutledge's ford, 34

Road from the fork to Rutledge's ford in Appomattox River above the mouth of Bush River, 66

NOTE: Entries for Sailors Creek are combined with Saylors Creek

Road from the Sand road into Craddock's road, 8

Road from the Church to Stocks Creek and continued to Sandy Creek, 8

Road from William Eckhols's road on Stocks Creek up to the ridge at the fork of Sandy Creek, 17

Road from Sandy Creek to the fork of the road to Jeneto, 55

Road from Sandy Creek to the Folly, 64

Road from Stocks Creek to Sandy Creek, 10, 31, 51

Road from Saylors/Sailors Creek to Sandy Creek, 19, 50, 64

Road from Sandy ford on Appamatox River to the main branch of Spring Creek, 31

Road from Sandy ford to Col. Randolph's mill and Vaughan's Creek, 36

Road from the mouth of Boush River below the mouth of Sandy River to Walker's road, 9

Road from Sandy River where Capt. Walker's old road crossed it to Bush River road, 37

Road from George Moore's to Sandy River/upper fork of Sandy River, 49, 54, 59, 64

Road from Sandy River to Nash's mill on Bush River, 59

Road from the race paths at Abraham Wawmock's to Sandy River bridge from thence to Bush River bridge and into the Buffelow road, 53

Road from Sailors Creek bridge to the fork of the road above Sandy River bridge to Bush River bridge and from thence to Buffelow road, 65

Road from the fork above Sandy River bridge to the fork above George Walker's, 65

Bridle way from Anthony Griffin's road to Sandy River Chapel, 22

Bridle way from the head of Snails Creek into Bush River road to Sandy River Chapel, 51

Road from the new road near Mallary's Creek to Sandy River road to the Church, 30

Road from Thomas Bevil's fork to Old Saponey ford, 31

Road from Flatt Creek to Sappony ford and Deep Creek lower bridge, 1

Road from the low grounds to Sappony ford, 13

Road from Neal's to Sappony road, 36

Road from Sawneys Creek into Randolph's road, 53

Old Rolling road from Randolph's road near his mill and thence across Vaughan's Creek at the old ford and then to continue the old way to the mouth of Sawneys Creek, 67

NOTE: Entries for Sailors Creek are combined with Saylors Creek

Road from Flatt Creek to the fork of Saylors Creek, 4

Mr. Walker's road from Saylors Creek to Crawford's and thence into Burton's road to the Court House, 8

Road from Stocks Creek to Saylors Creek, 10

Road from Saylors/Sailors Creek to Sandy Creek, 19, 50, 64

Road from Boush River to Saylers Creek, 20

Road from Boush River bridge across Saylors Creek to Walker's road, 21

Road from George Ham's into the road to Saylors Creek, 36

Road from Saylors Creek to Mr. Nash's road above Womawk's, 51

Road from Sailors Creek bridge to the fork of the road above Sandy River bridge to Bush River bridge and from thence to Buffelow road, 65

Road from or near Whitworth's in Saylors Creek road below the race paths, 29

Road from Joseph Ward's house into Sailors Creek road, 39, 40

Saylors Creek road turned at Charles Johnson's into the road from Crafford's to Dawson's race paths, 51

Road from Saylors Creek road to the road that goes to Crawford's house, 68

Road from Hudson's ford to the road to Rutledge's ford near the school house, 34

Road from Booker's mill to the fork at the school house, 63

Road by James Scotts house, 42

Road from the head of Little Flatt Creek to the fork road by Mr. Sherwin's plantation, 19

Road from Smacks Creek to Flatt Creek, 10

Road from Smax Creek to the river and Deep Creek bridge, 2

Road from Hall Creek to Smax Creek, 3

Road from Richard Booker's mill to Smax Creek, 13

Road round Mr. Smith's fence, 34

Road from Flatt Creek at Burton's bridge to Smith's Creek bridge at Winterham and from James Ferguson's to the Church, 64

NOTE: Variant spellings of Snails/Snales Creek are combined

Road from Watson's muster field along the ridge to the first branch of Snales Creek, 21

Road from Snails Creek to Lunenburg line, 33

Road from Snales Creek to the County line, 35

Road from Snales/Snails Creek across Mallary's Creek, 35, 61

Road from Snails Creek to George Moore's house and to Col. Richard Randolph's upper quarter, 37

Road from George Moore's to Snails Creek, 37, 61

Bridle way from the head of Snails Creek into Bush River road to Sandy River Chapel, 51

Road along the ridge from Moore's below Dejarnett's Smith Shop near head of Snails Creek, 58

Road from Flatt Creek/Flatt Creek bridge to Southall's/Southall's ordinary, 35, 47

Road near Southall's ordinary, 35

Road from Southall's to the Court House, 47

Road from Rockey Run to Spencer's Branch, 29

Road between Thomas Spencer's and Edmund Booker, Jr.'s into the main road from the Court House to Warwick near the south fork of Knibbs Creek, 44

Road from Spiner's/Spinner's Run to Wintercomake, 16, 62

Road from Spinner's to Dandy's race paths, 16

Fork of Spinner's road, 20

Road from Sandy ford on Appamatox River to the main branch of Spring Creek, 31

Road from Spring Creek to Lunenburg line, 35

Road to Capt. Stark's quarter, 1

Road near Capt. Starke's plantation, 12

Road from Dandy's race paths to Capt. Stark's new quarter, 16

Bridle way from George Steegall's house to the Church road, 67

Road from the White Oak on Flatt Creek to John Hurt's near the fork of Stocks Creek, 4

Road from the Church to Stocks Creek, 8, 33

Road from the Church to Stocks Creek and continued to Sandy Creek, 8

Road from Stocks Creek to Sandy Creek, 10, 31, 51

Road from Stocks Creek to Saylors Creek, 10

Road from Stocks Creek to the Court House, 10

Road from William Eckhols's road on Stocks Creek up to the ridge at the fork of Sandy Creek, 17

Road beginning above and near the mouth of Stocks Creek into the main road over Flatt Creek, 18

Road from Stocks Creek to Flatt Creek, 27

Road from Gillintine's/Gullington's to Stocks Creek, 32, 62

Francis Anderson's road (continued to Stocks Creek), 49

Road from Stock/Stocks/Stokes Creek to Samuel Major's, 49, 59, 62

Road to Stoker's bridge, 44

Road from Stoker's bridge to Jackson's road, 60

Stoker's road, 46

Road from Robert Stoker's to Clayborn's, 54

Road from Richard Stone's to the head of Buffelow-Bed Creek, 60

Road from Edmund Walker's house to Stony bridge, 63

Road from Stony bridge to Jeneto bridge, 64

Road over the Swethouse Creek below Abraham Jones's quarter to the main road, 1

Road from Namozain bridge to the Swett house Creek, 1

Road crossing Sweathouse Creek, 67

Road from Thomas Tabb's house to the road over Appomatox River bridge, 13

Road from Edward Jones's to cross West's Creek near Tally's Branch then between Wilkinson's quarter and Ward's to the Court House, 29

Road from Tanner's to Craddock's, 4

The road called Tanner's (see also William Belcher's road), 36

Road from Goode's bridge to Tanner's road thence to Anderson's road and to Thomas Jones's, 59

Road from the fork of Anderson's road down to Webster's commonly called Tanner's road, 67

Road from Thomas's to Main Nottoway, 20

Road from Thomas Anderson's house to Thomas's road, 18

Road from Thomas's road to Little Nottoway bridge, 60

Road from John Thomas's to the bridge from Nottoway and to ridge path near the race path, 20

Road from William Tinstall's house to the road leading to Appomatox, 24

Road from the County line between Tomahitton and the Birchen Swamps to the Chapel on Nottoway, (Road from Nottoway Chapel to Prince George County line) 4, 15, 59

Road from Bush River road at Beisley's path crossing the creek below Watson's mill thence into Watson's road below Tunstall's quarter, 37

Road from the head of Yarborrough's road up the ridge between Peters Creek and Whetstone to Tukaer's cart path then along the path up the ridge between Little Nottoway and Great Nottoway to the road near Degernett's, 50

Road from Appomatox River near Col. Richard Randolph's quarter up to Hill's fork on Vaughn's Creek, 17

Road from Sandy ford to Col. Randolph's mill and Vaughan's Creek, 36

Road from Vaughan's Creek to Randolph's road, 65

Old Rolling road from Randolph's road near his mill and thence across Vaughan's Creek at the old ford and then to continue the old way to the mouth of Sawneys Creek, 67

Bridle way from Robert Vaughan's to the Court House, 11

Road from Great Nottoway at Hampton Wade's to Nottoway bridge   59

Road from the mouth of Boush River below the mouth of Sandy River to Walker's road, 9

Road from Boush River bridge across Saylors Creek to Walker's road, 21

Road from Sandy River where Capt. Walker's old road crossed it to Bush River road, 37

Road from Edmund Walker's to the road by Nicholas Guilintine's, 61

Road from Edmund Walker's house to Stony bridge, 63

Road from George Walker's plantation to Buffalo River, 5

Road from the fork above Sandy River bridge to the fork above George Walker's, 65

Road from Mr. Walker's plantation into the road from Col. Randolph's quarters, 5

Mr. Walker's road from Saylors Creek to Crawford's and thence into Burton's road to the Court House, 8

Road from John Martin's in Mr. Walker's road, 9

Road from Flatt Creek bridge at Mayes's to the fork of Bush River road near Walters's road, 53, 56

Road from Ward's quarter to Anderson's road, 14

Road from Ward's quarter to the foot of the hills the other side Flatt Creek and Craddock's bridge, 14

Road from Edward Jones's to cross West's Creek near Tally's Branch then between Wilkinson's quarter and Ward's to the Court House, 29

Road through/round Benjamin Ward's plantation, 38

Road from Joseph Ward's house into Sailors Creek road, 39, 40

Road between Thomas Spencer's and Edmund Booker, Jr.'s into the main road from the Court House to Warwick near the south fork of Knibbs Creek, 44

Road from Goode's bridge to Warwick, 48

The first fork above the fork of the road to Warwick at Col. Cobbs' ordinary, 72

Road from Watson's muster field along the ridge to the first branch of Snales Creek, 21

Road from Mallery's Creek to the race paths at Watson's, 25, 61

Road from Bush River road at Beisley's path crossing the creek below Watson's mill thence into Watson's road below Tunstall's quarter, 37

Road from Watson's to the head of Bryery River, 59

Road from Watson's race ground to the fork below Daniel Jones's, 59

Road from Watson's race ground to Baldwin's/Baldwin's ordinary, 62, 71

Crossroads (fork?) where Watson's road comes to Flatt Creek, 20

Road from Bush River road at Beisley's path crossing the creek below Watson's mill thence into Watson's road below Tunstall's quarter, 37

The Parsons road from Watson's road into Irby's road, 47

Road from Watson's road to Nottoway Church, 59

Road from Capt. Watson's to Malerey's Creek, 47

Road from Capt. Watson's cart path to Mallery's Creek, 49

Roads from Capt. Watson's to the fork above George Moore's, 54

Road from William Watson's into the road at Brathwait's, 49

Road from the race paths at Abraham Wawmock's to Sandy River bridge from thence to Bush River bridge and into the Buffelow road, 53

Road from the fork of Anderson's road down to Webster's commonly called Tanner's road, 67

Road from John Robertson's ford into the road to Good's bridge below Peter Webster's, 45

Road from West Creek to the fork of the road near Capt. Peter Jones's quarter/Maj. Peter Jones's, 1, 43

Bridle way/road from West Creek to the Court House/Court House road, 2, 7, 15, 61, 64

Bridle way from West Creek to the Chapel on Nottoway, 2

Road from Thomas Bottom's on West Creek to the Old Ponds of Flatt Creek along or near the old Ridge path, 3

Road from the old pond to West Creek, 8

Road from the Church to West Creek, 13

Road from Charles Irby's house to West Creek, 15

Road from Nottoway to West Creek (Nottoway road to West Creek), 20, 42, 47, 48, 58

Road from West Creek to Buckskin, 20

Road from Flatt Creek to Appomatox crossing West Creek, 20

Road from West Creek to Bush River road, 22

Road from Cock's quarter to West Creek, 27

Road from Edward Jones's to cross West's Creek near Tally's Branch then between Wilkinson's quarter and Ward's to the Court House, 29

Road from Hudson's on West Creek down to Mann's road, 33

Road into the road leading to the bridge over West Creek, 40

Road from West Creek to the fork of the road near Maj. Peter Jones's, 43

Road from West Creek to Irby's road, 47

Road from Nottoway Church to West Creek, 55

Road from Nottoway Church along Court House road to West Creek, 59

Road from the cross road to West Creek, 61

Road from West Creek to Deep Creek, 61

West Creek road, 5, 13

Road from West Creek road just below Hudson's race paths down to Hugh Chambers' plantation, 36

Road from the Court House road to West Creek road below Hudson's race paths, 67

Road from the head of Yarborrough's road up the ridge between Peters Creek and Whetstone to Tukaer's cart path then along the path up the ridge between Little Nottoway and Great Nottoway to the road near Degernett's, 50

Road from the White Oak on Flatt Creek to John Hurt's near the fork of Stocks Creek, 4

Road from or near Whitworth's in Saylors Creek road below the race paths, 29

Road from Edward Jones's to cross West's Creek near Tally's Branch then between Wilkinson's quarter and Ward's to the Court House, 29

Road from Abram Bakers mill path to the upper Botton through Mr. Wimbush's and thence to the County line, 58

(Proposed) bridge where Winingham's road crosses Nottoway River, 35

Road beginning a little below John Winn's to Fisher's cart path and from thence to Jordan's bridge, 14

NOTE: All variant spellings of Winticomack (i.e., Wintercomake, etc.) are combined

Road from Capt. Peter Jones's fork to Wintercomake, 2

Road crossing over Wintercomake, 9

Road from Capt. Jones's quarter to Wintercomake, 11

Road from Spiner's/Spinner's Run to Wintercomake, 16, 62

Road from Wintocomake to Namaszeen/Namozain bridge, 16, 71

Road from Wintercomacke to Rockey Run, 29, 71

Road from James Coles's Spring Branch to Wintercomake, 29

Road from Green's road to Wintocomeck bridge, 62

Road from Flatt Creek at Burton's bridge to Smith's Creek bridge at Winterham and from James Ferguson's to the Church, 64

Road from Saylors Creek to Mr. Nash's road above Womawk's, 51

Road from Yarbrow's to Woody Creek crossing Irby's road, 20

Road from Richard Dennis' above Woody Creek, 70

Road from John Robertson's across Flatt Creek through the land of John Gibbs and Essex Worsham to the road to Good's bridge, 38

Road from Peter Wynn's road to William Jackson's new road at Bates's path, 44

Road from Yarbrow's to Woody Creek crossing Irby's road, 20

Road from William Yarbrough's house to Capt. Irby's road to the Court House, 29

Yarbrough's road crossing Little Nottoway, 41

Yarbrough's road, 45

Road from the head of Yarborrough's road up the ridge between Peters Creek and Whetstone to Tukaer's cart path then along the path up the ridge between Little Nottoway and Great Nottoway to the road near Degernett's, 50

www.ingramcontent.com/pod-product-compliance
Lightning Source LLC
Chambersburg PA
CBHW081148230426
43664CB00018B/2843